Liberal Protestantism and Science

LESLIE A. MURAY
Foreword by John B. Cobb, Jr.

Greenwood Guides to Science and Religion
Richard Olson, Series Editor

Greenwood Press
Westport, Connecticut • London

Library of Congress Cataloging-in-Publication Data

Muray, Leslie A.
  Liberal protestantism and science / Leslie A. Muray ; foreword by John B. Cobb, Jr.
      p.   cm. — (Greenwood guides to science and religion)
   Includes bibliographical references and index.
   ISBN 978–0–313–33701–7 (alk. paper)
      1. Religion and science.   2. Protestant churches—Doctrines.   I. Title.
   BL240.3.M865   2008
   261.5'5—dc22   2007036138

British Library Cataloguing in Publication Data is available.

Library of Congress Catalog Card Number: 2007036138
ISBN: 978–0–313–33701–7

First published in 2008

Greenwood Press, 88 Post Road West, Westport, CT 06881
An imprint of Greenwood Publishing Group, Inc.
www.greenwood.com

Printed in the United States of America

The paper used in this book complies with the
Permanent Paper Standard issued by the National
Information Standards Organization (Z39.48–1984).

10 9 8 7 6 5 4 3 2 1

To my friends and colleagues
Russ Pregeant
David Fedo
The late Joe Schneider
Laura Hubbard
and my alter ego
J. Ron Engel

Contents

Foreword

Those of us who have been related to theological education in what used to be the mainline churches, Episcopal, Presbyterian, United Church of Christ, Disciples, and United Methodist, find it hard to understand the now widely held public image of Protestantism in the United States. The same is true of the majority of Lutherans and American Baptists. Nowhere in our theological seminaries is there even a question raised as to whether we should respect and admire science and historical scholarship and adjust our teaching to their reliable findings and results. If Creationism and Intelligent Design are discussed at all, it is usually as illustrations of the wrong way of relating to science. This has been true for generations. It is hard for us to understand that so many people ignore this whole development and think and write as if no significant body of Christians had ever developed a positive relationship with science.

Furthermore, this tradition of friendly relations with science is not something new. In the late Middle Ages Christians laid the foundations of modern science. They pursued science in order to understand God better, and this motivation persisted into the eighteenth century. The church never objected. Of course, the transition from Aristotelian science to the modern world view occasioned some controversy. Most of it was among scientists. But in a period when theological heretics and "witches" were being slaughtered by the church, no scientist was killed, or even severely punished, for propounding scientific theories that radically transformed the philosophical and worldview basis of Christian theology.

In England in the eighteenth and nineteenth centuries, more science was done in the manse than in the university. As Muray points out, Darwin was a deist for most of his life, friend of clergy, and contributor to the church;

even after all the controversies his work engendered the Church of England bestowed on him its highest honor, burial in Westminster Cathedral. In the United States, when some Presbyterians rejected historical and scientific scholarship, they did not carry their denomination with them, but split off and founded another one. When some Protestants have tried to force their ideas on local school boards or state legislatures, the leaders of what were until recently the mainstream denominations could be counted on to oppose these efforts.

Those of us who have lived our entire lives in the church surrounded by Christians who are profoundly respectful of science find it hard to understand how people who seem otherwise well informed can ignore this mainstream history and identify Protestant Christianity only with what was until recently a fringe. I hope that Muray's brief sketch of the history of Liberal Protestantism's and its embrace of science will compel writers in the field to be at least a little more nuanced in their denunciation of "religion."

Perhaps some of these writers will recognize that among religious people there is always the danger of absolutizing present or past forms of their tradition and resisting all change. Perhaps they will recognize that similar tendencies exist among scientists and other scholars who often resist criticism of their assumptions and methods or even their cherished theories. Everyone recognizes that this kind of excess characterized Fascists and Communists and that many secular people give to their nations the sort of devotion that Christians believe should be given only to God. If we meet one another with the awareness that all of us need to be more critical of our traditions and ourselves, our discussions will be far more fruitful.

For me, as a participant in the "liberal" Protestant tradition that was until recently mainstream, such critical reflection about myself and my tradition is important. Clearly our invisibility today means that we have made some serious mistakes. Where did we go wrong? There is no consensus among us on this point, but I will express my personal views.

I believe that we went wrong by accepting the modern scientific worldview too uncritically. Prior to Darwin this meant accepting a dualistic view of a materialistic, mechanistic nature and a purely mental human soul. This is profoundly unbiblical, and in my view, profoundly wrong. It has intensified the anthropocentric tendencies already present in the Bible and the Christian tradition. It has destroyed the university created in the medieval period, centering in the humanistic liberal arts, and replaced it with multiple autonomous disciplines none of which relate effectively to the real issues now faced by humanity. And it bears considerable responsibility for the difficulties we still have in responding intelligently to the environmental crisis that threatens to destroy us all.

Please do not misunderstand me. I am not questioning that the sciences made great advances making use of this worldview. But in the twentieth century we discovered that it also blocked the understanding of new discoveries in physics. The reader of this book will discover that many Liberal Protestants have joined a few scientists in calling for replacement of the mechanistic paradigm with an organic one. We should have argued for this sooner and with greater insistence. We should do so now. This is *not* an attack on science. But it *is* a proposal for revising many scientific formulations that are shaped not only by the evidence but by a seventeenth-century metaphysics.

The one great crisis in the relation between science and religion has proved itself difficult to resolve chiefly because of the materialistic mechanistic model of nature. Given this model, the inclusion of human beings fully within nature, an inclusion that is rightly called for by evolutionary theory, has horrendous results. It means that we are asked to view ourselves as machines. I cannot believe that the scientists who press this view on us, intentionally or not, really adopt it for themselves. It was bad enough for religious people to accept this view of other animals, and these have paid, and continue to pay, a high price for our acquiescence. But scientists have no right to demand that religious people, or any others, adopt this view of human beings. To insist that we, scientists and religious people alike, are more than matter in motion is not to oppose *science*. It *is* to oppose a metaphysics to which too many scientists are committed. If we liberal Christians had made this conviction clearer to the general public, or even within our own congregations, perhaps we could not be so easily ignored today.

I am grateful for this book that makes so accessible to the general reader the reality of a major Christian tradition. The fault of this tradition in relation to science is far better understood as too ready an acceptance of whatever the "experts" say than any resistance to the authority of science in its proper sphere. Perhaps the current crisis of Liberal Protestantism will press us into speaking our distinctive contribution to the discussion with a louder voice, even if that sometimes offends our friends in the sciences.

John B. Cobb, Jr.

Series Foreword

For nearly 2,500 years, some conservative members of societies have expressed concern about the activities of those who sought to find a naturalistic explanation for natural phenomena. In 429 B.C.E., for example, the comic playwright, Aristophanes parodied Socrates as someone who studied the phenomena of the atmosphere, turning the awe-inspiring thunder which had seemed to express the wrath of Zeus into nothing but the farting of the clouds. Such actions, Aristophanes argued, were blasphemous and would undermine all tradition, law, and custom. Among early Christian spokespersons there were some, such as Tertullian, who also criticized those who sought to understand the natural world on the grounds that they "persist in applying their studies to a vain purpose, since they indulge their curiosity on natural objects, which they ought rather [direct] to their Creator and Governor."[1]

In the twentieth century, though a general distrust of science persisted among some conservative groups, the most intense opposition was reserved for the theory of evolution by natural selection. Typical of extreme antievolution comments is the following opinion offered by Judge Braswell, Dean of the Georgia Court of Appeals: "This monkey mythology of Darwin is the cause of permissiveness, promiscuity, pills, prophylactics, perversions, pregnancies, abortions, pornography, pollution, poisoning, and proliferation of crimes of all types."[2]

It can hardly be surprising that those committed to the study of natural phenomena responded to their denigrators in kind, accusing them of willful ignorance and of repressive behavior. Thus, when Galileo Galilei was warned against holding and teaching the Copernican system of astronomy as true, he wielded his brilliantly ironic pen and threw down a

gauntlet to religious authorities in an introductory letter "To the Discerning Reader" at the beginning of his great *Dialogue Concerning the Two Chief
World Systems*:

Several years Ago there was published in Rome a salutory edict which, in order to
obviate the dangerous tendencies of our age, imposed a seasonable silence upon
the Pythagorean [and Copernican] opinion that the earth moves. There were those
who impudently asserted that this decree had its origin, not in judicious inquiry,
but in passion none too well informed. Complaints were to be heard that advisors
who were totally unskilled at astronomical observations ought not to clip the wings
of reflective intellects by means of rash prohibitions.
    Upon hearing such carping insolence, my zeal could not be contained.[3]

No contemporary discerning reader could have missed Galileo's anger
and disdain for those he considered enemies of free scientific inquiry.
    Even more bitter than Galileo was Thomas Henry Huxley, often known
as "Darwin's bulldog." In 1860, after a famous confrontation with the
Anglican Bishop Samuel Wilberforce, Huxley bemoaned the persecution
suffered by many natural philosophers, but then he reflected that the
scientists were exacting their revenge:

Extinguished theologians lie about the cradle of every science as the strangled
snakes beside that of Hercules; and history records that whenever science and
orthodoxy have been fairly opposed, the latter has been forced to retire from the
lists, bleeding and crushed, if not annihilated; scotched if not slain.[4]

The impression left, considering these colorful complaints from both
sides is that science and religion must continually be at war with one
another. That view of the relation between science and religion was reinforced by Andrew Dickson White's *A History of the Warfare of Science
with Theology in Christendom*, which has seldom been out of print since it
was published as a two volume work in 1896. White's views have shaped
the lay understanding of science and religion interactions for more than a
century, but recent and more careful scholarship has shown that confronta-
tional stances do not represent the views of the overwhelming majority of
either scientific investigators or religious figures throughout history.
    One response among those who have wished to deny that conflict con-
stitutes the most frequent relationship between science and religion is to
claim that they cannot be in conflict because they address completely different human needs and therefore have nothing to do with one another.
This was the position of Immanuel Kant who insisted that the world of
natural phenomena, with its dependence on deterministic causality, is
fundamentally disjoint from the noumenal world of human choice and

morality, which constitutes the domain of religion. Much more recently, it was the position taken by Stephen Jay Gould in *Rocks of Ages: Science and Religion in the Fullness of Life* (1999). Gould writes:

I ... do not understand why the two enterprises should experience any conflict. Science tries to document the factual character of the natural world and to develop theories that coordinate and explain these facts. Religion, on the other hand, operates in the equally important, but utterly different realm of human purposes, meanings, and values.[5]

In order to capture the disjunction between science and religion, Gould enunciates a principle of "Non-overlapping magisterial," which he identifies as "a principle of respectful noninterference."[6]

In spite of the intense desire of those who wish to isolate science and religion from one another in order to protect the autonomy of one, the other, or both, there are many reasons to believe that theirs is ultimately an impossible task. One of the central questions addressed by many religions is what is the relationship between members of the human community and the natural world. This question is a central question addressed in "Genesis," for example. Any attempt to relate human and natural existence depends heavily on the understanding of nature that exists within a culture. So where nature is studied through scientific methods, scientific knowledge is unavoidably incorporated into religious thought. The need to understand "Genesis" in terms of the dominant understandings of nature thus gave rise to a tradition of scientifically informed commentaries on the six days of creation which constituted a major genre of Christian literature from the early days of Christianity through the Renaissance.

It is also widely understood that in relatively simple cultures—even those of early urban centers—there is a low level of cultural specialization, so economic, religious, and knowledge producing specialties are highly integrated. In Bronze Age Mesopotamia, for example, agricultural activities were governed both by knowledge of the physical conditions necessary for successful farming and by religious rituals associated with plowing, planting, irrigating, and harvesting. Thus religious practices and natural knowledge interacted in establishing the character and timing of farming activities.

Even in very complex industrial societies with high levels of specialization and division of labor, the various cultural specialties are never completely isolated from one another and they share many common values and assumptions. Given the linked nature of virtually all institutions in any culture it is the case that when either religious or scientific institutions change substantially, those changes are likely to produce pressures for change in the other. It was probably true, for example, that the attempts of pre-Socratic investigators of nature, with their emphasis on uniformities in the natural world and apparent examples of events systematically directed toward particular ends, made it difficult to sustain beliefs in the old pantheon of human-like and fundamentally capricious Olympian gods. But it is equally true that the attempts to understand nature promoted a new notion of the divine—a notion that was both monotheistic and transcendent, rather than polytheistic and immanent—and a notion that focused on both justice and intellect rather than power and passion. Thus early Greek natural philosophy undoubtedly played a role not simply in challenging, but also in transforming Greek religious sensibilities.

Transforming pressures do not always run from scientific to religious domains, moreover. During the Renaissance, there was a dramatic change among Christian intellectuals from one that focused on the contemplation of God's works to one that focused on the responsibility of the Christian for caring for his fellow humans. The active life of service to humankind, rather than the contemplative life of reflection on Gods character and works, now became the Christian ideal for many. As a consequence of this new focus on the active life, Renaissance intellectuals turned away from the then dominant Aristotelian view of science that saw the inability of theoretical sciences to change the world as a positive virtue. They replaced this understanding with a new view of natural knowledge, promoted in the writings of men such as Johann Andreae in Germany and Francis Bacon in England that viewed natural knowledge as significant only because it gave humankind the ability to manipulate the world to improve the quality of life. Natural knowledge would henceforth be prized by many because it conferred power over the natural world. Modern science thus took on a distinctly utilitarian shape at least in part in response to religious changes.

Neither the conflict model nor the claim of disjunction, then, accurately reflect the often intense and frequently supportive interactions between religious institutions, practices, ideas, and attitudes on the one hand, and scientific institutions, practices, ideas, and attitudes on the other. Without denying the existence of tensions, the primary goal of the volumes of this series is to explore the vast domain of mutually supportive and/or transformative interactions between scientific institutions, practices, and knowledge and religious institutions, practices, and beliefs. A second goal is to offer the opportunity to make comparisons across space, time, and cultural configuration. The series will cover the entire globe, most major faith traditions, hunter-gatherer societies in Africa and Oceania as well as advanced industrial societies in the West, and the span of time from classical antiquity to the present. Each volume will focus on a particular cultural tradition, a particular faith community, a particular time period, or a particular scientific domain, so that each reader can enter the fascinating story of science and religion interactions from a familiar perspective.

Furthermore, each volume will include not only a substantial narrative or interpretive core, but also a set of primary documents which will allow the reader to explore relevant evidence, an extensive annotated bibliography to lead the curious to reliable scholarship on the topic, and a chronology of events to help the reader keep track of the sequence of events involved and to relate them to major social and political occurrences.

So far I have used the words "science" and "religion" as if everyone knows and agrees about their meaning and as if they were equally appropriately applied across place and time. Neither of these assumptions is true. Science and religion are modern terms that reflect the way that we in the industrialized West organize our conceptual lives. Even in the modern West, what we mean by science and religion is likely to depend on our political orientation, our scholarly background, and the faith community that we belong to. Thus, for example, Marxists and Socialists tend to focus on the application of natural knowledge as the key element in defining science. According to the British Marxist scholar, Benjamin Farrington, "Science is the system of behavior by which man has acquired mastery of his environment. It has its origins in techniques ... in various activities by which man keeps body and soul together. Its source is experience, its aims, practical, its *only* test, that it works."[7] Many of those who study natural knowledge in preindustrial societies are also primarily interested in knowledge as it is used and are relatively open regarding the kind of entities posited by the developers of culturally specific natural knowledge systems or "local sciences." Thus, in his *Zapotec Science: Farming and Food in the Northern Sierra of Oaxaca*, Roberto González insists that

Zapotec farmers ... certainly practice science, as does any society whose members engage in subsistence activities. They hypothesize, they model problems, they experiment, they measure results, and they distribute knowledge among peers and to younger generations. But they typically proceed from markedly different premises—that is, from different conceptual bases—than their counterparts in industrialized societies.[8]

Among the "different premises" is the presumption of Zapotec scientists that unobservable spirit entities play a significant role in natural phenomena.

Those more committed to liberal pluralist society and to what anthropologists like González are inclined to identify as "cosmopolitan science," tend to focus on science as a source of objective or disinterested knowledge, disconnected from its uses. Moreover they generally reject the positing of unobservable entities, which they characterize as "supernatural." Thus, in an *Amicus Curiae* brief filed in connection with the 1986 supreme court case which tested Louisiana's law requiring the teaching of creation science along with evolution, for example, seventy-two Nobel Laureates, seventeen state academies of science and seven other scientific organizations argued that

[s]cience is devoted to formulating and testing naturalistic explanations for natural phenomena. It is a process for systematically collecting and recording data about the physical world, then categorizing and studying the collected data in an effort to infer the principles of nature that best explain the observed phenomena. Science is not equipped to evaluate supernatural explanations for our observations; without passing judgement on the truth or falsity of supernatural explanations, science leaves their consideration to the domain of religious faith.[9]

No reference whatsoever to uses appears in this definition. And its specific unwillingness to admit speculation regarding supernatural entities into science reflects a society in which cultural specialization has proceeded much farther than in the village farming communities of southern Mexico.

In a similar way, secular anthropologists and sociologists are inclined to define the key features of religion in a very different way than members of modern Christian faith communities. Anthropologists and sociologists focus on communal rituals and practices which accompany major collective and individual events—plowing, planting, harvesting, threshing, hunting, preparation for war (or peace), birth, the achievement of manhood or womanhood, marriage (in many cultures), childbirth, and death. Moreover, they tend to see the major consequence of religious practices as the intensification of social cohesion. Many Christians, on the other hand, view the primary goal of their religion as personal salvation, viewing society as at best a supportive structure and at worst, a distraction from their own private spiritual quest.

Thus, science and religion are far from uniformly understood. Moreover, they are modern Western constructs or categories whose applicability to the temporal and spatial "other" must always be justified and must always be understood as the imposition of modern ways of structuring institutions, behaviors, and beliefs on a context in which they could not have been categories understood by the actors involved. Nonetheless it does seem to us not simply permissible, but probably necessary to use these categories at the start of any attempt to understand how actors from other times and places interacted with the natural world and with their fellow humans. It may ultimately be possible for historians and anthropologists to understand the practices of persons distant in time and/or space in terms that those persons might use. But that process must begin by likening the actions of others to those that we understand from our own experience, even if the likenesses are inexact and in need of qualification.

The editors of this series have not imposed any particular definition of science or of religion on the authors, expecting that each author will develop either explicit or implicit definitions that are appropriate to their own scholarly approaches and to the topics that they have been assigned to cover.

Richard Olson

NOTES

1. Tertullian, 1896–1903. "Ad nationes." In Peter Holmes, trans., *The Anti-Nicene Fathers*, ed. Alexander Roberts and James Donaldson, Vol. 3 (New York: Charles Scribner's Sons), p. 133.

2. Christopher Toumey, *God's Own Scientists: Creationists in a Secular World* (New Brunswick, NJ: Rutgers University Press, 1994), p. 94.

3. Galileo Galilei, *Dialogue Concerning the Two Chief World Systems: Ptolemaic and Copernican* (Berkeley: University of California Press, 1953), p. 5.

4. James R. Moore, *The Post-Darwinian Controversies: A Study of the Protestant Struggle to Come to Terms with Darwin in Great Britain and America, 1870–1900* (Cambridge: Cambridge University Press, 1979), p. 60.

5. Stephen Jay Gould, *Rocks of Ages: Science and Religion in the Fullness of Life* (New York: The Ballantine Publishing Group, 1999), p. 4.

6. Ibid., p. 5.

7. Benjamin Farrington, *Greek Science* (Baltimore: Penguin Books, 1953).

8. Roberto Gonzáles, *Zapotec Science: Farming and Food in the Northern Sierra of Oaxaca* (Austin: University of Texas Press, 2001), p. 3.

9. *72 Nobel Laureates, 17 State Academies of Science and Seven Other Scientific Organizations. Amicus Curiae* Brief in support of Appelles Don Aguilard et al. vs. Edwin Edwards in his official capacity as Governor of Louisiana et al. (1986), p. 24.

Acknowledgments

I would like to begin the acknowledgment of the many people who made this book possible with my parents, Remus F. and Marianna Muray, deceased in 1994 and 1996. Not only did they bring me into the world and raise me in the difficult world of a totalitarian regime, a revolution, of being refugees and immigrants, they also modeled, in different ways, a love of nature and the integration of religion and science.

I cannot express adequately the debt of gratitude I feel toward John B. Cobb, Jr. He was my professor and mentor the at the Claremont School of Theology and Claremont Graduate University, drawing the best out of me to this day. He is an inspiration, a colleague, and a deeply caring friend. His influence can be felt on every page of this book.

I want to thank all of my friends at Highlands Institute for American Religious Thought and the Center for Process Studies in Claremont, CA, for their friendship and encouragement over the years. I especially want to thank Jon Taylor, Cedric Heppler, Nancy Howell, Ed Towne, Jennifer Jesse, Vaughan McTernan, Frederick Ferré and Barbara Ferré, Jerry and Sue Stone, Del and Nancy Brown, Howard and Rita Radest, George Allan, and Charley Hardwick.

Another special note of thanks goes to David Ray Griffin, another of my former professors at Claremont, friend and colleague, who suggested to the series editor, Richard Olson, that I be the author of this volume. My gratitude goes to Dick Olson for selecting me to write this volume, for his helpful comments, and for our jovial get-togethers. To Kevin Downing, my editor at Greenwood Publishing, patient and a delight to work with. I hope we get to see a Red Sox-Tiger game together soon. And my thanks go to my copyeditors, Umananda K. and Kakoli Sajwan, and to my production manager, Anoop Chaturvedi.

In a different vein, I want to thank Steve Nelson, former All-Pro linebacker and at the time Head Football Coach and Athletic Director at Curry College, for insisting that I go to the hospital after what turned out to be a heart attack in June, 2004. He probably saved my life. I also want to express heartfelt thanks to my cardiologist, Dr. Scott Lutch, for his inspirational care.

The first two chapters of this book were written during the summer of 2005. I would like to thank my colleagues in English, Susan Peterson and the late Joe Schneider, for their bibliographical help as well as their insights into the Romantic movement described in chapter two.

The rest of the book was written while I was on sabbatical in my native Hungary July, 2006 to January, 2007. Living in an apartment on the banks of the Danube was very conducive to writing in the mornings and occasionally after long walks along the Danube in the evenings.

I want to express my deepest thanks to János ("Jimmy") Kelemen, my host in Hungary, Chair of the Department of Philosophy and Director of the Institute of Philosophy at Eötvös Lorand University, Budapest. It was at his invitation that I spent the sabbatical in Hungary and taught "American Philosophy" and "Environmental Ethics" at the afore mentioned university. Jozsef Nagy, Researcher in the Department of Philosophy at Eötvös Lorand University, was a virtual teaching assistant whose gracious help knew no bounds.

Erzsébet Nagy was like a sister to me, especially at times when I felt like a stranger in the country of my birth. Her boundless energy is an inspiration. She helped me navigate through some difficult times and circumstances and steered me through several wonderful Hungarian libraries. She is the embodiment of the Jewish version of the kind of religious liberalism described in this book.

I want to express my thanks to people who graciously invited me to lecture on parts of this book: to Gábor Karsai, President of the Gate of Dharma Buddhist College at the time, for inviting me to give a lecture on "Whitehead and Science" to the Hungarian and Central European Whitehead Society; to Mihály Toth, Professor of Philosophy at Péter Pázmány University at Piliscsaba and one of the most creative contributors to the religion–science dialogue in Hungary today, for inviting me to give the same lecture to his Process Philosophy Seminar; to László Szabó, Chair of the Philosophy of Science Colloquium, Eötvös Lorand University, for inviting to speak to that group on "Whiteheadian Process Philosophy and Science"; and to Béla Mester who invited me to lecture to the Institute of Philosophical Research, Hungarian Academy of Sciences on "The Enlightenment and Romanticism," Chapter 2 of this book. Finally, for the

friendship and stimulating work of Gábor Kovács on the Hungarian political thinker István Bibó and for the friendship and inspirational leadership
of a group of searchers on the part of Ildikó Németh.

I would be remiss not to mention three wonderful friends and their
significant others whom I met during my sabbatical: István Kovács, Zoltán
Penczel, and Péter Ficsor; and their respective significant others, Kata,
Aniko, and Gabriella. I treasure their friendship. Our conversations about
this book as well as other topics, their perceptive questions, deep insights,
and sense of humor were among the best parts of my sabbatical and made
the writing easier. I cannot mention them without mentioning Timea and
Évike at the Ipoly kávéház. They contributed to our discussions and our
celebration of life.

I want to express appreciation to my eight-year-old, three-legged cat,
Sasha, for the great source of joy and companionship she provides. I wonder if she could imagine how much I missed her during my six-month
absence!

I have to mention an aunt and numerous cousins with whom I was
reunited for the first time in fifty years; some of them, I was meeting for
the first time. While they are too numerous to mention by name, I do
want to thank Laci and Aranka, their daughter Petra, Gyuri and Kati, and
Annus who encouraged me to take time each day during my emotional
reunions at my father's village to work on this book. Most of the part on
the neoorthodox interlude, especially the section on Reinhold Niebuhr,
was written at this time. A very special thank you to my aunt Marcsi who
lived within our household during the first eight years of my life and with
whom I have been reunited after forty-nine years! It was a delight to meet
her children, my cousins about whom I had heard so much, Toncsi, Feri,
and Laci. And another special note of to my cousins Pisti and Maria, Pisti,
Jr., and Andi, my cousins in Budapest and to my cousin Laci who lives in
the Great Plain (Alföld).

My appreciation goes to Joe Murphy ("Murph"), with whom I became
good friends in Budapest, truly a warmhearted human being. Amazing
that he, an ardent Yankee fan, and I, a devout Red Sox fan, are such good
friends. And to Krisztina and our friends at the Sip Sarok Kavé.

A special note of gratitude goes to my colleagues, in the Humanities
Department, especially to Dean David Fedo, until recently Academic Dean
and Vice President, President Ken Quigley, and to the Board of Trustees
of Curry College for granting me the sabbatical that enabled me to finish
this project.

My appreciation goes to John Cobb and my colleagues Hedi Ben Aicha,
John Hill, and Russ Pregeant for reading the manuscript and making
helpful suggestions.

The book is dedicated to my friends and colleagues at Curry College, Russ Pregeant, David Fedo, the late Joe Schneider, Laura Hubbard, and to my alter ego, J. Ron Engel, Professor Emeritus of Social Ethics at Meadville Lombard Theological School.

Chapter 1

◆

Introduction

If we look at popular depictions and the conventional stereotype of the relationship between science and religion, whether in the media or such popular plays and movies as *Inherit the Wind*, the dominant picture we get is that of eternal warfare (an allusion to Andrew Dickson White's *A History of the Warfare of Science with Theology in Christendom*; White, President of Cornell University, helped to popularize the notion that the only possible relationship between religion and science was warfare). Moreover, this is the only possible option in the relationship that is presented, it is the dominant way the story is told.

In my view, until fairly recently, i.e., the last twenty-five years, at least a part of the public was fairly well read and aware of the history I am telling. And although today we have some major daily newspapers that have reporters assigned to the religion beat, we do not seem to have figures of such eminence as the late Associated Press reporter George W. Cornell who in his popular *The Untamed God*, highly touted by Billy Graham on the cover, alluded to positive ways of seeing the science–religion relationship.

Yet, the story of the warfare between religion and science is hardly the whole story. Often untold is the long history of the radical, enthusiastic, unequivocal embrace of modern science (and the secularity that usually comes with it) on the part of Liberal Protestant church bodies and theologians (as well as progressive Roman Catholics, Jews, and Muslims) that is an integral part of the history of modernity and postmodernity. This book is an endeavor to tell that story.

The story has some premodern antecedents. Although the typical popular stereotype of faith is that it is a great leap into the dark regardless of the evidence, hence irrational, there is a long tradition, one that, it is possible to claim, was dominant until at least the Reformation, that saw faith and reason not only as compatible but also as complementary. While this confidence in reason expressed itself in a deductive mode of reasoning that unquestioningly assumed certain things to be true (unlike modern inductive, scientific reasoning that created a hypothesis, tested it, and let the evidence dictate the conclusions) and then reasoned about them, it nevertheless sowed the seeds that led to the emergence of modern science. Scientists like Galileo who came into conflict with the church were products of church universities that instilled confidence in the capacities of reason and a questioning attitude.

Moreover, there is another time-honored tradition within the Christian tradition, from the time of the Church Fathers, beginning with Clement of Alexandria and Origen at the end of the second century, first half of the third century down to today's liberals, namely "the Two Books" concept of God's revelation. According to this concept, God has revealed the divine self in two primordial ways, two different but equally valid books: the "Book of Scripture" and the "Book of Nature"; the latter is understood through the exercise of reason and examination of the evidence.

Having said that, I want to set the stage for telling of story of Liberal Protestantism by using the typologies of Ian G. Barbour and John F. Haught concerning the relationship between science and religion, both of which I find particularly useful (Barbour 1990, 3–30, 1997, 77–105, 2000; Haught 1995). The two typologies parallel each other.

Barbour delineates four ways of relating science and religion: conflict, independence, dialogue, and integration (Barbour 1997, 77–105). The conflict model is espoused by advocates of scientific materialism (the terminology is somewhat inexact: since not all conflict model advocates are full blown materialists, it might be more accurate to call them scientific naturalists. But then not all who identify themselves as scientific naturalists identify themselves as advocates of the conflict model) who see the universe in reductionistic terms: what we can know is restricted to that which we can know through sense experience and events can be explained with reference to their lowest common denominator, the laws of physics and chemistry (Barbour 1997, 78). Such a view does not allow much room for religion.

The conflict model is also espoused by adherents of biblical literalism (Barbour 1997, 82–84). Advocates of this view reject either modern science (at least selectively) if science contradicts literal readings of the Bible, which is viewed as inerrant, or, more recently, attempt to find scientific evidence for such literalism, as we can see among advocates of "creation science" and the more fundamentalist devotees of "Intelligent Design."

The dominant popular image of the relationship between science and religion, as mentioned above, is that of the conflict model. The only possible alternative positions characteristic of this image are scientific materialism and biblical literalism, with the advocates of each position doing much to reinforce this public perception.

The second model of Barbour's typology is the independence model. According to this view, science and religion, though equally valid, use different methods, speak different languages, and deal with different realities. This is the position of those evangelical and conservative Christians who accept modern science. It has been the position of Protestant theologians influenced by existentialism. It is also the position of some scientists who are also religious. As we shall see, it is a possible position within the tradition of Liberal Protestantism and was adopted as such during the neo-orthodox interlude in the Liberal Protestant tradition in the twentieth century (Barbour 1997, 84–89).

The third of Barbour's models for the relationship between science and religion is that of dialogue. Although distinct in their concerns and as academic disciplines, science and religion need to enter into mutually fruitful dialogue about "limit questions" and about methodological parallels (Barbour 1997, 90–93). Limit questions, such as basic trust that life is meaningful, joy, anxiety, and being confronted with death that arise at horizons of human experience provide the context within which religious questions arise, questions that do not escape even the most rigorous scientists (Barbour 1997, 92). With regard to methodological parallels, there has been an increased recognition that, contrary to science's claims to pure objectivity, there is an element of subjectivity involved, that the scientist participates in what is observed as well as the results of observation. Conversely, there has also been a growing recognition that elements of both objectivity and subjectivity are involved in religion (Barbour 1997, 94).

Another dimension of the dialogue model is provided by a nature-centered spirituality. This can be seen in the works of a wide variety of writers: Rachel Carson; the novels of Annie Dillard; the creation-centered spirituality of Matthew Fox that sees the sacred in nature; Thomas Berry and Brian Swimme who narrate history in a science-based non-anthropocentric way, that is to say, non-human centered way; the work of David Bohm and Fritjof Capra both of whom drew connections between religions and quantum physics; and numerous others (Barbour 1997, 95–98). All of them want to affirm the sacredness of the non-human world in contrast to the dominant anthropocentric Western tradition that has deprecated the world of nature. We shall have occasion to deal at greater length with some of these in a subsequent chapter on ecotheology.

The fourth model is that of integration. One example of this is provided by contemporary forms of "natural theology" that begin with scientific evidence, then argue for evidence of some sort of purpose or design, as in the arguments to the effect that the dynamics of the universe were pre-disposed, in terms of probabilities, for the development of life and human beings (the Anthropic Principle) and that such order in the universe suggests a Designer (not to be confused with contemporary arguments about Intelligent Design) (Barbour 1997, 98–100). Another is found in theologies of nature that, with a concern for our ecological crisis, reinterprets inherited doctrines in light of science, something that, once again, we shall have occasion to encounter again at greater length in a later chapter. Finally, there are those who pursue a systematic synthesis of science and religion, something that has been attempted in Liberal Protestantism throughout the modern era down to today (Barbour 1997, 103–105).

John F. Haught's typology of the relationship between science and religion is very similar to Barbour's. His models are: conflict, contrast, contact, and confirmation (Haught 1995). The conflict model holds to the notion of irreconcilable differences between science and religion. It is a view shared by scientific materialists, scientific skeptics, biblical literalists, and non-fundamentalists who hold modern science responsible for the pervasive sense of meaninglessness in the modern world (Haught 1995, 9–12). Haught's conflict model is identical with Barbour's.

In similar fashion, his contrast model resembles Barbour's independence model. Rejecting the conflict model, scientists and theologians who adhere to the contrast model maintain that science and religion deal with distinct, separate spheres. As Haught points out, it is an ingenious way to get out of the impasse of the conflict model (Haught 1995, 14–15).

The third model is the contact model, which parallels Barbour's dialogue model. The following quote provides an excellent summary of this position: " . . . it is convinced that, without in any way interfering with scientists' own proper methods, religious faith can flourish alongside of science in such a way as to co-produce with it a joint meaning that is more illuminating than either can provide on its own" (Haught 1995, 18). Like Barbour, he points to the contemporary questioning of the *absolute* objectivity and the growing recognition that there is a subjective element in scientific knowing, that both science and religion use imaginative metaphors, that science is not detached, passionless endeavor, and that there are elements of social construction involved in the scientific enterprise (Haught 1995, 19–21).

Haught maintains that a critical realist epistemology (theory of knowledge) can serve as a bridge in the dialogue or contact between science and religion. This is not to be confused with naïve realism, which understands our understanding and knowledge of the world to be a "copy" of the way the world really is. Rather for critical realism, while there is a real, objective world, perception and knowledge of that world always involves a degree of subjectivity (Haught 1995, 20). Thus, in answer to the age-old question of whether the tree falling in the forest is making a noise if there is no one to hear it, the critical realist would maintain that indeed the tree is falling and that it is making a noise. However, she/he would add that if there were several hearers present, what was heard would depend on the subjective perception of the hearer.

Haught's fourth model is confirmation, once again quite similar this time to Barbour's integration model. Haught maintains that "I call this fourth approach 'confirmation,' a term equivalent to 'strengthening' or 'supporting,' because it holds that religion, when carefully purged of idolatrous implications, fully endorses and even undergirds the scientific effort to make sense of the universe" (Haught 1995, 21). Even with the distinctiveness of science and religion, there is a unity to truth. Moreover, if religion deals with the fundamental trust, faith, that life is meaningful and worth living, there is an element of that basic trust in science in presupposing the intelligibility and rationality of the universe, in seeing its quest for truth as worthwhile (Haught 1995, 22– 25).

I have used the typologies of Ian G. Barbour and John F. Haught to show that, contrary to the popular image and conventional stereotype, there is a long history of variety in the way religion and science have related to each other. In terms of their respective typologies, the most typical of these espoused by Liberal Protestantism's radical, enthusiastic, unequivocal embrace of modern science (and its concomitant secularity) are contact and integration-confirmation.

The rest of this book is devoted to telling the story of Liberal Protestantism's radical embrace of modern science (and its concomitant secularity). In the following chapter, I set the stage further by delineating some characteristics of the Enlightenment: confidence in reason, inductive, scientific reasoning; seeing reason as the defining characteristic of being human; a questioning attitude toward all authority; individualism; and confidence in progress. I treat Romanticism as a reaction against the rationalism of the Enlightenment, particularly in seeing feelings, the emotions as the defining characteristic of being human. I trace the roots of Liberal Protestantism to both the Enlightenment and Romanticism.

Chapter 3 looks at various forms of Liberal Protestantism's embrace of modern science in the form of Darwinism in the nineteenth century. The first section deals briefly with such representative thinkers as Lyman Abbott, Henry Drummond, and John Fiske, all of whom embraced some form of the notion that God creates through the evolutionary process. The next section describes how this Liberal Protestant embrace of science took the form of Social Darwinism, especially in Josiah Strong. The next section looks at Darwinism's impact on the Social Gospel while the final section deals with the Anglican embrace of Darwinism as typified by the collection of essays in the volume *Lux Mundi*, particularly the work of Charles Gore.

The following chapter treats the convergence of science and religion in the Liberal Protestantism of the late nineteenth, early twentieth centuries. While other schools will be mentioned, the "Chicago School" will be the primary point of reference. The neo- orthodox interlude will receive brief treatment since there is little interaction between science and religion, except in the work of Karl Heim whose work will be the prime focus.

In Chapter 5, the first section describes early attempts to revive the dialogue in 1960s. This will partly focus on the work of Daniel Day Williams, Harold K. Schilling, and Ian G. Barbour. The beginnings of *Zygon* and the work of Ralph Wendell Burhoe are also highlighted. The next section deals with the further continuance of dialogue and the beginnings of integration (again) in the subsequent works of Barbour, Birch, Cauthen, and Cobb. A variety of forms of integration typified by works published in *Zygon*, by such theologians as Phil Hefner, Ted Peters (both Lutheran) and Karl Peters (Unitarian Universalist), and in the works of such groups as the Institute on Religion in an Age of Science (I.R.A.S.), Center for Theology and the Natural Sciences (C.T.N.S.), and groups affiliated with the Templeton Foundation (the Metanexus) are examined. I conclude the chapter by considering process theologians such as Griffin, Haught, McDaniel, Howell, Keller, Pederson, Case-Winter, and Oord.

Chapter 6 looks at how concerns with the degradation of the environment have been an offshoot of the Liberal Protestant embrace of modern science. The works of John B. Cobb, Jr., Nancy R. Howell, Sallie McFague, Jay McDaniel, and Paul Santmire receive special attention. The final chapter provides a summary.

Epistemological issues, which provide perhaps the greatest source of division between religion and science, will be addressed throughout the course of the book.

Chapter 2

⌒

The Enlightenment and Romanticism

THE ENLIGHTENMENT

The Age of Science and the Enlightenment, the Age of Reason have much in common. Chronologically, historically, they are often lumped together (Livingston 1997, 5). The historical dating of these periods is rather arbitrary, with a great deal of overlap. For example, one could maintain that the high point of the Age of Science (historians of science often use "Age of Science" to refer to the nineteenth century, using "Scientific Revolution" to refer to the seventeenth century; I am using the designation "Age of Science" as the more typical designation in philosophy, history, and world civilization textbooks) was 1600–1700, of the Enlightenment 1700–1800, of Romanticism 1750–1850. But one could no less easily maintain, with a great deal of plausibility, that the Enlightenment lasted until at least 1850, Romanticism until the late nineteenth century. In fact, one could make the claim that the Age of Science, the Age of Reason, and Romanticism have continued to today. In my view, we are at the very least dealing with the legacies of each of these eras. I also want to distinguish them from each other in order to preserve the distinctiveness of each era.

What distinguishes the Age of Science are not just the great scientific discoveries of that era—Sir Isaac Newton and the laws of gravity, Harvey and the circulation of blood, Galileo—but, perhaps even more importantly, the rise of inductive scientific reasoning. This rationality was not the kind of premodern deductive reasoning that assumed certain things to be true and then reasoned why that was so. Rather, it created hypotheses, tested them, and then drew conclusions based on the evidence. In the broader philosophical and cultural arena, the emphasis on this kind of reasoning led to a sense that it was legitimate to question authority, in fact to question everything—whether it be the divine right of kings, the authority of the Bible, of the Pope, or of the church, Roman Catholic or Protestant (Livingston 1997, 67; Randall 1976, 261–271).

One of the key characteristics of the Enlightenment was seeing rationality as constitutive of the meaning being human. This kind of rationality was the inductive, unfettered questioning the Age of Reason adopted from the Age of Science (Barbour 1997, 38–39; Livingston 1997, 6–7; Randall 1976, 261–271). The importance of the Enlightenment's questioning attitude toward all authority and its impact on religious tradition cannot be emphasized enough (Livingston 1997, 6–7). Sometimes, especially in France, it took a strongly anti-ecclesiastical and anticlerical form— epitomized by Voltaire's (1694–1778) slogan "Ecraser l'infâme!" ("Crush or break up the infamy!") (Livingston 1997, 26) and the execution of numerous clergy, primarily Roman Catholic priests, during the French Revolution.

Previous ages (with the exception of the venerated ancients of the Classical Age) were seen as ages of darkness, ages of superstition, the age of the childhood of the human race, of human immaturity, much as they had been seen during the Renaissance. Humanity had finally reached the stage of maturity and adulthood, the Age of Reason. Through the exercise of inductive, scientific reasoning, humans had a historical vocation to shed the authority of the ignorant, superstitious, stifling past, and thus be liberated to be truly themselves.

Another feature of the Age of Science that the Enlightenment appropriated was its Newtonian (named after Sir Isaac Newton [1642–1727]) mechanistic view of the world. Typified by the model of the world as a clock, the non-human natural world was seen as a lifeless, inert machine made up of disposable parts (Barbour 1997, 38–39; Randall 1976, 254–279). Newton's mechanistic worldview was wedded to the dualism of the philosopher and mathematician René Descartes (1596–1650). Cartesian (after Descartes' name in Latin) dualism, the bifurcation of reality into irreconcilable opposites, saw mind and matter as separate and distinct entities. What was ultimately real was the human mind, the defining characteristic of being human ("Cogito, ergo sum." "I think, therefore I am."); the body was a mere unfeeling, unthinking machine-like, nature-like appendage to the mind. Moreover, the function of the human mind was no longer to discover the nature of reality writ large in the universe, as was the case in the view of Ancient/Classical and Medieval thinkers, but to interpret and impose order on that world. In the Newtonian- Cartesian view of world, it is only human beings who can think and feel and are therefore the only creatures of intrinsic value (Ferré 1996; Randall 1976, 239–242).

Quite typical of the Enlightenment's view of the world was the deism of many of its exponents. If the world was a marvelous, orderly machine that functioned according to the regularity of its own laws that could be discovered through inductive reasoning, it was reasonable to surmise that there was an orderer, a creator, a clockmaker who set the machine, the clock in motion, a clock which was perfectly capable of operating on its own, although the Clockmaker could intervene to make any necessary repairs (Barbour 1997, 239–242).

In a fashion similar to the Renaissance and in contrast to the communal emphasis of previous ages, the Enlightenment highlighted the importance of the individual. The inherent dignity of the individual and the need to safeguard it by political institutions and social arrangements was stressed. The political community was seen as an aggregate of individuals who contracted to form a political community of their own design.

No less an important and typical feature of the Enlightenment was its confidence in progress. We have already seen hints of this in the self-understanding of the great thinkers of that era who saw themselves as living in the Age of Reason, a new age of the maturity of the human race finally potentially free of the ignorance and dead weight of the suffocating past. If the inherent rationality of humans was unleashed, the potential of human beings had no limits (Livingston 1997, 8–9; Randall 1976, 381–385). For some, this progress was more or less automatic. Prometheus, the figure in Greek mythology who had stolen fire from the gods and thus made civilization possible, became the symbol of humanity and its full potentiality.

Profoundly tied to the previous characteristics of the Enlightenment was an advocacy of secularization. The words "secular," "secularity," "secularism," "secularization" have their roots in the Latin word "saeculum," which means "this world" or "this age." Once again in a manner reminiscent of the Renaissance, the great representative thinkers of the Enlightenment embraced this world in a radical fashion—in contrast to what they saw as the otherworldliness of the Middle Ages, which they saw as evidence of immaturity and ignorance. They embraced the world on its terms, in its own integrity. Part of this radical embrace of the world was the espousal of the *process* of secularization, the progressive removal of ideas and institutions from the dominance of religion (Berger 1967). In philosophy and theology, we can see a similar phenomenon in deism. Although most of the thinkers of the Enlightenment, for all their criticisms of traditional beliefs and institutional religion, still wanted to preserve their faith in God, the role of that God in the world was increasingly diminished. As mentioned previously, for deists, while God could intervene in the world for repairs, the world was perfectly capable of operating on its own terms. Explanations for the causation of events were in increasingly naturalistic terms with no reference to God. Eventually, God disappeared from the picture entirely. In the words of Laplace when Napoleon asked him what he thought of God: "Sire, I had no need of that hypothesis!" (Barbour 1997, 35).

There was also a drive to separate politics from the dominance of religion. This was seen in efforts to institutionalize religious tolerance and in efforts to separate church and state. This meant the "disestablishment" of religion, minimally meaning no official state religion, more expansively, granting equal rights to non-Christians and non-believers and not privileging religion (Randall 1976, 372–376). In France, as mentioned previously, this took a very anti-ecclesiastical and anti-clerical form, especially during the French Revolution.

One of the most important driving forces in the Enlightenment's radical embrace of the world and advocacy of secularization was the memory of the religious wars of the sixteenth and seventeenth centuries. Much of Europe had been devastated by thirty years of war between Protestants and Roman Catholics in the 1500s, with another Thirty Year War in the 1600s. France had endured nearly thirty years of civil war in the second half of the sixteenth century. England had its religious civil war in the seventeenth century. This led the thinkers of the Age of Science and the Enlightenment—those that remained at least nominally religious—to the question of how could they remain committed Christians yet be tolerant and avoid the nonsense of the religious wars? Related to this issue was the question of how could one be a contemporary Christian and accept the findings of modern science (still one of the vexing questions today)? What could one know with absolute certainty and yet be tolerant? The paradigmatic answer came from René Descartes: the only thing one could know with absolute certainty is oneself thinking!

As one might expect, there were a variety of responses to these pivotal features of the Enlightenment on the part of philosophers, theologians, and religious communities. One of these responses was deism itself: it certainly provided an answer to how to one might be a contemporary human being, embracing modern science, its methods, and the attendant progress of the human race, and retain one's faith in God!

A prototypical example of this deist response is Thomas Jefferson's (1743–1826) the *Jefferson Bible*. Jefferson edited the New Testament, leaving in only those parts that he thought were plausible to a modern human being. Thus, he left out the miracles and anything else with supernatural connotations. He arranged New Testament material chronologically to read like a biography, highlighting the teachings that dealt with how humans ought to relate to one another. This was done by a person who, although a deist and one who expressed sympathy with Unitarianism, remained a member of the Church of England, of the Episcopal Church, as the American branch of the Anglican Church was called after the colonies gained their independence (Jefferson 1989).

There are numerous other deists who provide good illustrations of the deist response to the Enlightenment. It was typical for the deists to maintain that *rational religion* and *Christianity* pointed to the same universal truth. For example, Matthew Tindal (1655–1733), in his *Christianity as Old as Creation*, maintained that the Bible was not a unique revelation but a particular manifestation of a universal truth. In France, while his polemic against traditional Christianity was at times virulent, Voltaire (1694–1778) did espouse a form of rationalist deist religion.

Another prototypical response is provided by another of the great figures of the Enlightenment, Immanuel Kant (1724–1804). Kant developed his epistemology, at least in part, as a response to the Scottish philosopher David Hume (1711–1776). Following John Locke, Hume maintained that the human mind was a "tabula rasa," a "blank slate" on which sense impressions were made. For the Scottish philosopher, experience was limited to sense experience and it was the only reliable means to attain knowledge. The school of philosophical thought that claims that the only way to know is through experience is called *empiricism*, which came into conflict with the *rationalism* of the Enlightenment. Identifying experience with sense experience became a distinctive feature of British empiricism and would provide the way modern science defines the word empirical (Barbour 1997, 44). Modern science uses the word "empirical" to mean that which we can know, namely that which can observed, putting the emphasis on the sense of sight.

Kant attempted to resolve the tension between empiricism and rationalism. While agreeing with Hume that there is no knowledge apart from experience, he also maintained that the human mind was not merely a passive recipient of sense data but was active in organizing, ordering, and interpreting the world and our experience of it. The world could not be known in itself but only as interpreted by the human mind.

Thus, for Kant, the empirical method of modern science is valid; it is one way to attain knowledge. However, he also highlighted the importance of subjectivity, freedom, and moral choice. And it is in our sense of moral obligation, in "practical reason," that we find the proper place of religion. Although, as a representative figure of the Enlightenment, he attempts to build morality "within the limits of reason alone," i.e., on strictly rational grounds with no recourse to external authority, such as revelation, he did claim that God and the notion of immortality are foundations and guarantors of moral order (Barbour 1966, 74–78, 1997, 45–47).

In espousing the idea of two different ways of knowing, with each having its separate realm and function, Kant opened the path for one of the ways of reconciling religion and science. He was a forerunner of Barbour's independence and Haught's contrast models.

It is time for us to consider the responses of religious institutions and religious thinkers and theologians more directly tied to religious institutions to the Enlightenment. Once again, there were a variety of responses ranging from hostility to defensiveness to embrace. Bearing a considerable amount of the Enlightenment's critique of religion, Roman Catholicism, particularly in France, was largely hostile and defensive. Anglican Bishop Joseph Butler (1692–1752) represented a more positive but defensive response as he attempted to defend revealed religion (vs. the "natural" religion of the Enlightenment) in his *Analogy of Religion, Natural and Revealed* (1736) His defense rested on an analogy between natural theology, which endeavors to set forth about God through reason alone, and revelation. Butler maintained that, contrary to the claims of the thinkers of the Enlightenment that the realm of nature was one of pure order, evidence from which pointed to order and reason, nature was full of ambiguities. He readily admitted that scripture was no less full of ambiguities. But amidst all the ambiguities, scripture provided no less evidence for the existence of God than did nature (Barbour 1997, 36– 37).

Much closer to the position of the deists is the Anglican philosopher/theologian William Paley (1743–1805). In his *Natural Theology* (1802), Paley argued for a variation of the traditional argument from design for the existence of God. That is to say, the order of the universe suggests a cosmic orderer, an Intelligent Designer. Paley used the analogy of the clock and clockmaker and, his favorite, the example of the structure of the eye working toward the purpose of seeing, as paradigms of the order of nature from which we can infer the existence of God (Barbour 1997, 51).

On the Protestant side, one of the clearest examples of divergent responses to the Enlightenment can be seen in the conflict between Congregationalism and Unitarianism in the late 1700s and first half of the 1800s, especially in the United States, most particularly in New England. The Puritanism of Congregationalism, for the most part, represented the traditionalist Protestant rejection of the Enlightenment—although one dimension of that Puritan tradition, seeing God as the source of all knowledge, including modern scientific knowledge, embraced modern science. With their questioning of the inherited tradition, especially the doctrine of the Trinity, the divinity of Christ, and the meaning and authority of Scripture, the Unitarians adopted the rationalism of the Enlightenment as the final arbiter of truth claims. Unitarians are among the best examples of a religious community's radical accommodation to the Enlightenment.

Although it is the part of the topic of another book in this series, I do feel the need to mention a parallel response in Judaism. The development of

Reform Judaism, to be sure in no small measure prompted by Emancipation, granting Jews the right to join civil society if they chose to do so, was in part a response to the Enlightenment. Reform Judaism wrestled and continues to wrestle with the question of what it means to be a modern human being and a believing, practicing Jew at the same time. In doing so, it adopted Enlightenment style rationalism as the final arbiter of truth.

ROMANTICISM

Scholars treat Romanticism either as part of (Livingston 1997, 83–84) or a reaction to the Enlightenment (Randall 1976, 399–400). While both of these views have plausibility, I would maintain that it is more helpful and accurate to adhere to the latter view if one is to uphold the integrity and distinctiveness of both the Enlightenment and Romanticism. I hope that my characterizations of the Enlightenment above and of Romanticism below provide my rationale for this claim.

The historical roots of Romanticism reside in the religious movement known as Pietism. Founded by Philip Jakob Spener (1635–1705), a Lutheran pastor, Pietism originated in the German speaking parts of Europe (there was no unified country of Germany at the time). Initially, Pietism was not so much a protest against the rationalism of the Enlightenment but rather against the formalism, rationalism, and scholasticism of the Lutheranism of the 1600s and early 1700s. For Pietism, authentic religion was not a matter of adherence to correct beliefs and regular participation in the liturgical life of the church. True religion was a matter of the "heart," of genuine and deep emotion and feeling, rather than the "head." It emphasized the experience of "regeneration," of "being born again," a deep and dramatic experience of Christ's love that turned one's life around. The revivalist tradition, which came to dominate the American frontier of the nineteenth century, had its origins in Pietism. It was very influential in the Great Awakening in the American colonies in the 1740s and in the ministries of John (1703– 1791) and Charles Wesley (1707–1788), the founders of the Methodist Church. The Great Awakening and Wesleyanism can be seen as expressions of Pietism (Barbour 1997, 41– 42).

In this context, I need to mention one of the key figures of the Great Awakening and one of America's greatest native born theologians, Jonathan Edwards (1703–1758). A staunch defender of Calvinism, he synthesized Calvinism with Newtonian physics, Pietism, "a sense of the heart," and practical ethical concerns. He anticipated and influenced such subsequent American philosophical and theological movements as New England Theology, American pragmatism, and neo-orthodoxy (H. Richard Niebuhr). (Although there seems to be no causal connection with either Pietism or Romanticism, and in fact chronologically Hasidism emerged

Figure 2.1 John Wesley (Courtesy Library of Congress)

before Enlightenment rationalism influenced Reformed Judaism, I see a remarkable parallel between Pietism and Hasidism. The historical origins of each are virtually simultaneous, late seventeenth, early eighteenth centuries. Hasidism, like Pietism, emphasized religion not as matter of the head, of rituals or rules but of the heart. It rebelled against formalism and the authority of the rabbis. Like Romanticism, it saw God in the world of nature and emphasized the immanence, the intimate closeness of God to creation). Rooted in Pietism, Romanticism became a much wider cultural movement, influencing the church but moving largely outside its boundaries.

Without abandoning the questioning of authority and the attitude of rebelliousness toward the authority of the past (although some Romantics were nostalgic about the Middle Ages), Romanticism saw the Enlightenment's emphasis on reason as rather arid and abstract. Consequently, as a reaction to this distinctive emphasis of the Enlightenment, Romanticism saw the defining characteristic of being human in terms of feelings and emotions rather than reason (There are scholars who maintain, in keeping with the notion that Romanticism is best viewed as part of the Enlightenment, that Romanticism attempted to hold reason and emotion in balance (Livingston 1997, 83–84). Although this argument has merit and would thus facilitate seeing romanticism as a precursor of radical empiricism, a contention with which I would agree, with which I shall deal at length later, nevertheless, in my view it blurs the distinctive emphases of the Enlightenment and Romanticism). Humans are creatures of the heart rather than the head. When Robin Williams tells Matt Damon, "Follow your heart, you'll be all right!" in *Good Will Hunting*, he is in effect stating the most fundamental feature of Romanticism. Part and parcel of the Romantics' focus on emotion and feelings is a concomitant emphasis on the concrete and particular rather than the abstract and universal emphasized by the great thinkers of the Enlightenment (Barzun 1976, xv; Livingston 1997, 83–84; Randall 1976, 399–403; Welch 1972, 52–55; Whitehead 1953, 75–94).

While this is the greatest point of contrast between the Enlightenment and Romanticism and is usually treated as such in scholarly literature, I would like to highlight several others. The second of these is the Enlightenment's proclivity to treat the non-human natural world in mechanistic terms. In contrast, the Romantics extolled the non-human natural world, relished its wonders. They saw that world not as a lifeless, inert machine strictly of instrumental value but took an organic view that saw nature as living, creative, of intrinsic value. The relationship to nature so emphasized by the Romantics (Wordsworth, Whitman, Emerson, Thoreau, and other New England Transcendentalists, the Hungarian Petôfi readily come to mind as does Mary Shelley's *Frankenstein*, one interpretation of which is that it is a protest against the Enlightenment and its practical consequences that ensued in the Industrial Revolution) went hand in hand with their emphasis on feelings (Livingston 1997, 84–85; Randall 1976, 417–423; Whitehead 1956, 75–94). In fact, they saw a correspondence between feelings and the non-human natural world that was felt (Abrams 1953, 65–68).

In keeping with their view of nature as organic, living, and creative, the Romantics tended to be *pantheists*, seeing God as the totality of things, or more accurately identifying God with the world of nature. For them, the sacred permeated all living things, expressing what can be construed as a modern form of animism, the presence of an animating spirit in all things. Unlike the distant God of the deists for whom there was precious little left to do, for the Romantics, God was immanent if not identical with nature. Although, there are some exceptions, most Romantics did not have much use for institutionalized religion and in effect adhered to a religion of nature.

Another point of contrast between the great figures of the Enlightenment and the Romantics concerns the latter's emphasis on the community. To be sure, both the Age of Reason and Romanticism extolled and celebrated the individual (Randall 1976, 415–417). However, the Enlightenment tended to do so in an abstract way while Romanticism emphasized the concreteness and particularity of feelings and passions. Moreover, the Romantics extolled the community (including the nation on the part of some of them) no less than the individual. Unlike the thinkers of the Enlightenment who saw political community as an aggregate of individuals who contracted to form a political community of their own design, the Romantics saw the political community in organic terms. Individuals, in their distinctiveness, were interrelated, interdependent, constitutive parts of these organic communities. While Romanticism did not use the term, in my view it anticipated the notion of the "individual-in-community," an idea integral to the thought of such diverse thinkers as Shailer Mathews, Nicholas Berdyaev, Alfred North Whitehead, and Bernard E. Meland.

Romanticism shared the Enlightenment's confidence in the potentialities of the human race. The figure of Prometheus became no less the symbol and prototype of modern humanity for the Romantics than he had been for the thinkers of the Enlightenment—exemplified by Percy Shelley's "Prometheus Unbound." In spite of this confidence, at least some of the Romantics were aware of the ambiguities of human feelings and potentialities; the full title of Mary Shelley's *Frankenstein* is *Frankenstein or the Modern Prometheus* (Shelley 1992, 19).

The Romantics' attitude toward secularization, hardly uniform, was complex. On the one hand, not having much use for the institutional and organizational aspects of religion, they promoted the progressive removal of ideas and institutions from the dominance of religious institutions. However, in another sense, they resisted the division, religious and secular, of the sacred and the profane. After all, if the world is permeated by the sacred, if God is immanent or identical with the world, how can one make such an arbitrary division?

The responses of the churches to Romanticism varied from hostility and a defense of the role of reason to influence and appropriation of its major tenets. New England Transcendentalism, which, along with the work of Walt Whitman, is the best example of American Romanticism, began as a rebellion against what it considered to be the formalism, traditionalism, and rationalism of Unitarianism.

The clearest example of the impact of Romanticism and of a positive response to it on the part of a church theologian was the theology of Friedrich Schleiermacher (1768–1834), often called "the father of modern Protestantism," "the father of liberal Protestantism." Both a pastor and a professor, he taught at the University of Halle and at the University of Berlin. The influence of Pietism is readily apparent in his theology. Adopting the notion that feelings characterize the meaning of being human, he maintained that religion is not a matter of beliefs, doctrines, rituals, reason, or morality but of feelings. More specifically, what comprises religious experience is feeling, namely "the feeling of absolute dependence." The feeling of absolute dependence is "God- consciousness," manifest fully in Jesus of Nazareth (Livingston 1997; Schleiermacher 1994).

By emphasizing religious experience as a particular dimension of human experience and as the distinguishing feature of religion, Schleiermacher, with Kant, paved the way for the independence and contrast models of liberal Protestantism's responses to modern science. Emphasizing experience as he did, he also paved the way for the more dialogical and integrationist embrace of modern science anticipating the broader and deeper definition of experience that we shall see in various forms of radical empiricism.

As we shall see in ensuing chapters, liberal Protestantism is a child of both the Enlightenment and Romanticism. Some of its adherents have espoused the rationalism of the Enlightenment, others Romanticism's focus on feelings and emotions. Still others have attempted to synthesize the two, not seeing reason and emotion as antithetical but different dimensions of the broad spectrum of experience.

In this regard, it is paradoxical that, at least in part, Protestant evangelicalism, including its fundamentalist expression, and Protestant liberalism have common roots in the revivalist, pietistic tradition of the late seventeenth, eighteenth, and early nineteenth centuries.

Chapter 3

The Darwinian Controversy and Liberal Protestantism

THE LIBERAL EMBRACE OF DARWINISM

It goes without saying that, from the time of the publication of *On the Origins of Species* in 1859, the focal point of the conventional stereotype about the eternal warfare between religion and science has been Charles Darwin and the theory of evolution. Nothing draws the ire of fundamentalists, few things attract the admiration of scientists like the theory evolution.

While this chapter is devoted to the Liberal Protestant embrace of Darwinism, a few comments about Darwin himself and his attitude toward religion are in order. Contrary to the conventional stereotype of a consistent, if not militant atheist, early in his life the British scientist contemplated studying for the ordained ministry (Phipps 2002, 1–15). For much of his life, he espoused a form of theism similar to deism (Phipps 2002, 75–77, 166). Consistent in his skepticism about revealed religion and the divinity of Christ throughout most of his life, in his later years he professed agnosticism, i.e., not knowing with certainty one way or the other about the existence of God (Phipps 2002, 142–155). Throughout the course of his life, he maintained his friendship and correspondence with such liberal clergy as Charles Kingsley and contributed generously to church organizations and missionary societies (Phipps 2002, 75, 148–155).

One aspect of Darwin's theory of evolution had important forerunners. The prevalent geological theory about God's creation of new species was *catastrophism*, the idea that the creation of new species occurs in-between a series of great upheavals and cataclysms. Scientists like James Hutton (1795) and most especially Charles Lyell in his *Principles of Geology* made a major breakthrough in advocating *Uniformitarianism*. This is the view that instead of the cataclysms described above, new species evolve and grow, through the operation of natural causes, slowly, over a long period of time, although it needs to be pointed that for Lyell this did not suggest evolution until the end of his life. Lyell's point of view became foundational for Darwin (Barbour 1997, 50; Livingston 1997, 250–253).

Before I turn to Liberal Protestantism's enthusiastic embrace of evolution, some comments are in order about why the Darwinian theory of evolution has seemed and seems to be so threatening. Some, conservative Christians and secular thinkers alike, see it in the threat to a literal reading of the Bible, especially of Creation. While, to be sure, this has merit and historical substantiation, it is nevertheless insufficient by itself to explain some of the visceral reactions to Darwinism—except as part and parcel of modernity's quest for absolute certainty! Non-literal and allegorical interpretations of the Bible were common from at least the time of Pharisaic Judaism, if not the earliest oral tradition that went into the formation of Hebrew Scripture. In spite of the vehement reaction of literalists and in fact also against historical critical readings of the Bible, such non-literal, historical critical treatments of scripture were not at all uncommon in Darwin's day.

Darwin's notion of natural selection as the engine that drives the evolutionary process poses a visceral threat at several levels as well as to the sense of absolute certainty. For one, there is the claim—and the evidence to back that claim—that the complex forms of life emerged from the simple, that in the case of human beings, we emerged from non-humans. This kinship between humans and non-humans posed and continues to pose a threat to the inherited anthropocentrism, human centeredness, of the Western tradition and the way it was internalized by nineteenth-century Victorian culture. Instead of being superior to the animals, to the brutes, humans were now alleged to have descended from non-humans. This seemed to reduce humanity's special place in the universe, to dislodge any sense of the special dignity of human beings. It took away any sense of mystery about human existence, about the sacred. Additionally, the notion of randomness of natural selection advocated by later Neo-Darwinists undermined and continues to undermine any sense of teleology in the universe—not only in the sense of an overarching purpose but specific purposes on the part of organisms.

Bishop William Wilberforce voiced these concerns in a caricatured way, thus reinforcing the conventional stereotype for the ages when, in a debate with Thomas Huxley in 1860, he asked Huxley on which side of his family, his mother's or his father's, was he descendent from the apes. To be historically accurate, Wilberforce, trying to be funny and reflecting his era's Victorian glorification of women, asked Huxley if the latter would be willing to trace his ancestry to an ape on his grandmother's side

(Livingston 1997, 256). According to legend, Huxley replied that if he had a choice of either the ape or Wilberforce, who in Huxley's estimation was being ridiculous in involving himself in scientific discussions, he would gladly choose the ape (Livingston 1997, 256)!

In spite of the convenient image that Wilberforce's one-liner in the first version of the story provides for the conventional stereotype, it would be historically inaccurate to leave the matter there. Within a short period of time after this incident, Liberal Protestants were articulating their espousal of Darwinian evolution both in academic and in channels more accessible to the public.

I have selected some representative figures of the Liberal Protestantism of the late nineteenth, early twentieth centuries to illustrate the Liberal Protestant enthusiastic embrace of the theory of evolution. The first of these thinkers is the Scottish lay theologian Henry Drummond (1851–1897).

Drummond was a layperson and a scientist. For him, "modernism," i.e., the acceptance of the modern world, including modern science and the use of historical critical methods to study the Bible, was combined with a personal pietism. Spirit is present throughout creation, that is to say throughout the evolutionary process. Since that is the case, humans emerged in the course of the cosmic evolutionary process; there is no need to postulate special divine intervention in the creation of humans or a special creation of the human soul (Barbour 1997, 68). The Laws of the Spirit are identical with the Laws of Nature, the supernatural with the natural (Drummond 1888, vi, 75–94).

Drummond maintained, in contrast to Darwin, that all living things have two basic functions: "Nutrition," which is the foundation for the "Struggle for Life" and "Reproduction," which for Drummond is synonymous with the "Struggle for the Life of Others." The Scottish thinker saw the place of the "Struggle for Life" as getting rid of the unfit and imperfect; without it, advance and progress would not be possible. However, the "Struggle for the Life of Others" is steadily triumphing over the "Struggle for Life," with the imminent Kingdom of Love being announced by science itself (Livingston 1997, 262–263). Thus, the path of progress and the path of altruism are identical and "evolution is nothing but the involution of Love" (Drummond 1894, 46; Livingston 1997, 263).

As we shall see with the other Liberal Protestant theologians we shall consider, Henry Drummond identifies two characteristics in his enthusiastic embrace of Darwinian evolution. The first of these has to do with the creative evolutionary process itself serving as one of the "Books of God," making (revealing) God and God's ways accessible to human reason. That is to say, the evolutionary process itself is a "pointer" to God and how to live in harmony with God and God's creation. The other characteristic has to do with allowing room for some sort of teleology, some sort of purposiveness, not only the particular purposes of particular organisms but of the ultimate triumph of the Kingdom of Love. As we shall see, such confidence in an Ultimate or Final End was not a characteristic of liberals, particularly after the second half of the twentieth century.

The prototype of Liberal Protestantism's radical and enthusiastic embrace of evolution is Lyman Abbott (1835–1922). Abbott had succeeded Henry Ward Beecher at Plymouth Church in Brooklyn, New York. Working from such a prestigious pulpit, he exerted considerable influence not only among academics but also the educated public as well (today's would be "public intellectuals" can be justifiably jealous!). Abbott stood within the tradition of Christian apologetics, that is to say, that trajectory within the Christian tradition that uses the best available resources in the culture to make faith intelligible to those both outside and inside of it. As such, Christian apologetics takes a positive attitude toward and makes use of reason and its most adequate expression. For Abbott, evolution provided the most adequate intellectual framework for making Christianity intelligible for educated, thinking women and men of his day (Livingston 1997, 263).

As was the case with Drummond, Abbott identified the spiritual with the laws of nature (Livingston 1997, 263). For him, the evolutionary process was never arbitrary but worked according to the laws manifest in the union of nature and spirit. Thus, the laws of the evolutionary process entail the evolution of the complex forms of life from the simple (Livingston 1997, 263). Abbott writes that "God's work, we evolutionists believe, is the work of progressive change—a change from lower to a higher condition; from a simpler to a more complex condition" (Abbott 1897, 21). Moreover, he claims that "the process of growth is produced by forces that lie in the phenomena themselves . . ." and that God dwells in nature fashioning it according to His will by vital processes from within, not by mechanical processes from without" (Abbott 1897, 21).

The implications of appropriating evolution for the development of "the New Theology," to use Abbott's terminology, meant getting rid of ancient and medieval tyrannical, hierarchical, feudal models of God in which God is removed and distant from the world and unaffected by it, ruling by divine fiat through external interventions. In contrast, he advocates a view that sees God's relation to the world as totally immanent and recast completely in evolutionary terms. He maintains that God's only way of doing things is the way of evolution, of growth and development:

. . . that He resides in the world of nature and in the world of men; that there are no laws of nature which are not the laws of God's own being; that there are no forces of nature, that there is only one divine infinite source . . . ; that there are not occasional or exceptional theophanies, but that all nature is and all life is one great

Figure 3.1 Lyman Abbott (Courtesy Library of Congress)

theophany; that there are not occasional interventions that bear witness to the presence of God, but that life is itself a perpetual witness to His presence. (Abbott 1897, 9–10)

In rejecting hierarchical, monarchical models of God in which God is unaffected by the world, Abbott is anticipating the concerns of contemporary feminists and process theologians.

The long time pastor of Brooklyn's very prestigious Plymouth Church also applied evolution to the Bible and to the idea of revelation. God's revelation is always progressive. Abbott writes that " ... revelation is not a final statement of truth, crystallized into dogma, but a gradual and progressive unveiling of the mind that it may see truth clearly and receive it vitally" (Abbott 1892, 25). Moreover, he asserts that:

The Bible is not fossilized truth in an amber Book; it is a seed which vitalizes the soil into which it is cast; window through which the light of dawning day enters the quickened mind; a voice commanding humanity to look forward and to go forward; a prophet who bids men seek their golden age in the future, not in the past. (Abbott 1892, 25)

Consistent with his progressive/processive view of revelation, he maintains that the Bible itself is the product of evolution, of natural selection (Abbott 1892, 26–67). In his progressive/processive view of revelation, "revelation is the unveiling in human consciousness of that which God wrote in the human soul when he made it" (Abbott 1892, 116). Abbott states that:

...in the heart of man God has written his message, his inviolable law and his merciful redemption, because he has made the heart of man akin to the heart of God. Revelation is the upspringing of this life of law and love, of righteousness and mercy, under the influence of God's personal presence and power. (Abbott 1892, 116)

This poses a fundamental choice between the "Old Theology" and Abbott's understanding of the "New Theology":

To the Old Theology, God, as a great infinite Caesar ruling the world, has framed certain statutes and given them to us, and we must obey them, or come into collision with him and suffer the threatened penalties. To the New Theology, he has made man after his own image and written his own nature in the human consciousness and in human love, and then has interpreted by the mouth of his prophets what he has written in the hearts of his children. (Abbott 1892, 116–117)

Thus, although God discloses the divine self and the way for humans to live, that revelation is not revelation until humans beings receive it—in their very human, partial, imperfect, fragmentary ways.

Abbott extends his progressive/processive view of revelation and the Bible to an optimistic view of human beings and history. He rejects the traditional doctrines of the Fall and Redemption. If humanity has been evolving from the simple forms of life to the complex, this means that, however slowly, human beings have been progressing spiritually (Abbott 1892, 206). Thus, sin is "regression" from a higher, more developed spirituality to the lower, and in keeping with Victorian anthropocentric sensibilities, "animal" forms of life from which we have evolved (Abbott 1892, 227).

Abbott typifies liberal theology's emphasis on the continuity between the human and the divine. If there is continuity rather than an unbridgeable gulf between God and humans, then there is no difference in essential nature between God and human beings (Abbott 1897, 188–189). Moreover, the difference between Jesus Christ and other human beings is a matter of degree and not of kind (Abbott 1897, 73).

Abbott saw the ultimate goal of history as the evolution of divinity out of humanity (Livingston 1997, 264). The following is an eloquent summary of his evolutionary theology:

Incarnation is the indwelling of God in a unique man, in order that all men may come to be at one with God. . . . Finally: religion is not a creed, long or short, nor a ceremonial, complex or simple, nor a life more or less perfectly conformed to an external law; it is the life of God in the soul of man, re-creating the individual; through the individual constituting a church; and by the church transforming human society into the kingdom of God. (Abbott 1892, 257–258)

Abbott's "theology of an evolutionist" is a form of evolutionary theism that anticipates contemporary process thought and other recent expression of evolutionary theism. It is also a classical prototype of the Liberal Protestant view that God creates the universe through the evolutionary process, a continuing creation in which human beings are the junior partners.

I maintained earlier that Abbott appropriated Darwinism into his theology. He drew a distinction between his own "theology of an evolutionist" and Darwinism, which he identified with Social Darwinism's advocacy of untrammeled "laissez-faire" capitalism (Livingston 1997, 263), an issue I shall discuss in greater detail below and in the next section on Social Darwinism. Suffice it to say for our purposes, that in spite of Abbott's rejection of Darwinism because he saw it as synonymous with Social Darwinism, he nevertheless used such ideas as the "survival of the fittest," which he otherwise abhorred, to make sense of his understanding of the "evolution of Christianity," to explain why some interpretations of the inherited tradition were not fit to survive (Abbott 1892).

Although he was not a professional theologian, one of the most interesting examples of the reconciliation between evolution and theism is provided by John Fiske (1842–1901). His fame has been revived in recent years as a member of the Metaphysical Club, which met in Boston 1873–1875 on a regular basis for scientific and philosophical discussions. The Metaphysical Club included, besides Fiske, such future luminaries as Charles Sanders Peirce, William James, and Oliver Wendell Holmes, all very much influenced by the less well-known Chauncey Wright. William James credited Wright with coining the term "pragmatism" for a distinctive school of philosophy.

For Fiske, the conflict between religion and science has arisen unnecessarily, namely on account of identifying God as external, outside of the universe, an identification carried out by many scientists and theologians alike. If God is seen as the indwelling spirit of the universe who/that is the immanent driving force of the creative evolutionary process, there need not be a conflict between religion and science. Fiske maintains that such Church Fathers as Athanasius anticipated modern evolutionary science and evolutionary theism with an emphasis on God as the indwelling spirit of the universe who/that animates all things (Fiske 1887, 87–110).

The one-time member of the Metaphysical Club rejects the traditional argument from design, the idea that the order of the universe suggests a cosmic orderer or designer. Intriguingly anticipating process thought and other contemporary schools of thought that, as we shall see, attempt to redefine God's power as limited, Fiske critiques the traditional argument from design for espousing a form of divine determinism that collapses on the shoals of the theodicy problem, the problem of evil, or more precisely, why is there evil in the world if God is *all good and all powerful* (Fiske 1887, 121–127)? Moreover, he takes particular aim at the kind of mechanistic determinism that had been dominant since the seventeenth century, typified for him in the work of Paley. He states that Paley's simile of the watch needs to be replaced by the simile of the flower:

The universe is not a machine, but an organism, with an indwelling principle of life. It was not made, but it has grown. (Fiske 1887, 131)

To Fiske, the idea of evolution has been the source of the greatest revolution in human thinking (Fiske 1887, 131–132). The laws of evolution are applicable to all phenomenon—including, as we shall see, human societies. The pervasiveness of the laws of evolution " . . . means that the universe as a whole is thrilling in every fibre with Life—not indeed, life in the usual restricted sense, but life in a general sense" (Fiske 1887, 149). Anticipating, as we shall see, Whiteheadian process thought, he states that "the distinction, once deemed absolute between the living and the not living is converted into a relative distinction; and Life as manifested in the organism is seen to be only a specialized form of the Universal Life" (Fiske 1887, 149).

In a manner reminiscent of Yoda teaching Luke Skywalker in *The Empire Returns*, Fiske writes: "All is quivering energy" (Fiske 1887, 150). And behind all phenomena, behind all forms of energy, is an Energy. He calls this energy Force but he considers this a term that is inadequate and unenlightening (Fiske 1887, 151). Consequently, he also calls it the Power manifest in all phenomena, which is the source of all matter but is itself non-material: it is psychical (Fiske 1887, 152–153). Thus, in Fiske's view, "from the crudest polytheism we have thus, by a slow evolution, arrived at pure monotheism, the recognition of the eternal God indwelling in the universe, in whom we live and move and have our being" (Fiske 1887, 155). Although he describes God in psychical terms, he is concerned that we do not project limitation and weakness onto God. In this regard, he diverges from Whiteheadian process thought as well as most other contemporary understandings of God, according to which God's power is limited (Fiske 1887, 155–156).

Although, as we have seen, Fiske rejects traditional forms of teleology, including that of Paley, he not only affirms the purposiveness of particular actions but also maintains that "...the Darwinian theory of natural selection...when thoroughly understood it will be found to replace as much teleology as it destroys" (Fiske 1887, 158). And he does think that, in at least a loose sense, evolution has an ultimate goal, pointing to "...a discernible dramatic tendency, a clearly marked progress of events toward a mighty goal" (Fiske 1887, 159). He states that "...in the story of the evolution of life upon the surface of our earth, where alone we are able to compass the phenomena, we see all things working together, through countless ages of toil and trouble, toward one glorious consummation" (Fiske 1887, 159). For Fiske, "the glorious consummation toward which organic evolution is tending is the production of the highest and most perfect psychical life" (Fiske 1887, 160).

Although humanity has a kinship with the non-human natural world and emerged from it, Fiske does not escape the anthropocentrism of his era: "We see Man still the crown and glory of the universe and the chief object of divine care..." (Fiske 1887, 165). And in a fashion we have seen in Lyman Abbott and typical of the Victorian context of the late nineteenth in the United States, he saw a constant struggle in human beings between their "higher" and their "lower," "animal" natures (Fiske 1887, 165).

The following are Fiske's concluding words about God:

The infinite and eternal Power that is manifested in very pulsation of the universe is none other than the living God.... But of some things we may feel sure. Humanity is not a mere local incident in an endless and aimless series of cosmical changes. The events of the universe are not the work of chance, neither are they the outcome of blind necessity. Practically there is a purpose in the world whereof it is our highest duty to learn the lesson, however well or ill we may fare in rendering a scientific account of it. When from the dawn of life we see all things working together toward the evolution of the highest spiritual attributes of Man, we know, however, the word may stumble in which we try to say it, that God is in the deepest sense a moral Being. The everlasting source of phenomena is none other than the infinite Power that makes for righteousness. Thou canst not by searching find Him out; yet put thy trust in Him, and against thee the gates of hell shall not prevail: for there is neither wisdom nor understanding nor counsel against the Eternal. (Fiske 1887, 166–167)

Like Lyman Abbott, John Fiske, in spite of the fact that he was not a professional theologian, is a prototypical example of the Liberal Protestant (his work did find resonance among Unitarians and Liberal Protestants) embrace of evolution. He provides another example typifying Liberal Protestantism's affirmation that God created the world through the evolutionary process, which is a "continuing creation" in which humans have a distinctive part to play. Fiske's "evolutionary theism," as we have seen, anticipates a number of contemporary evolutionary theists.

A profoundly ambiguous dimension of Fiske's thought is the influence of Herbert Spencer. We have seen the influence of Spencer, to which Fiske alludes often, in his allusions to God as the Force or Power behind all phenomena. Fiske was also very much influenced by Spencer's application of the laws of evolution to social evolution, the evolution of human societies. What is not apparent in Fiske's writings about God and evolution is the influence of Spencer as one of "the fathers of Social Darwinism."

As a Social Darwinist, Spencer helped popularize the notion of "the survival of the fittest." Spencer applied this idea from biological evolution to social evolution. Thus, in a world of scarce food supplies and resources, it is only the strong, "the fit," those that deserve to survive who survive. In socio-politico-economic terms, that means "laisser- faire," "leaving things alone," that is, no governmental interference in either economic or political life. To establish government programs to help the poor goes against the ways of nature (Randall 1976, 331, 444–445, 500, 506, 603).

This sanctification of extreme, untrammeled laissez-faire capitalism on Darwinian grounds was coupled with an affirmation of the superiority of the Anglo-Saxon race and Rudyard Kipling's racist call "to take up the white man's burden." For John Fiske, a historian, this took the form of calling for an active American foreign policy which was not afraid to spread the benefits of democracy and aggressively pursue American interests abroad.

SOCIAL DARWINISM

We have looked at some basic features of Social Darwinism. In this section, we shall focus on Josiah Strong (1847–1916) as a prototype of a Liberal Protestant clergyperson who appropriates Darwinism into his/her basic theological outlook and winds up an ambiguous advocate of Social Darwinism. The influence of Josiah Strong cannot be minimized: he was a friend of President Theodore Roosevelt, who saw to it in 1900 that Strong was introduced to Admiral Mahan, the advocate of American sea power. America's arrival as a world power acquired a new religious legitimation.

Strong was the long time (1886–1898) General Secretary of the Evangelical Alliance, a coalition of a variety of missionary groups. He served churches in Ohio. His most famous and influential work was *Our Country: Its Possible Future and Present Crisis* (1885). The book is motivated by a concern for missionary activity in the West, in the cities, and abroad. He also has a concern for ending racial conflict. The answer to humans ills resides in bringing all people to Christ. And, in this endeavor, "the Anglo- Saxon race" has a special role.

In Strong's estimation there are numerous perils threatening American society. These are (in the late nineteenth, early twentieth centuries): Mormonism; Socialism; Intemperance; Wealth; "Romanism;" Urbanization; and Immigration. Cities are a particular target of his polemic: they are points of attraction for "inferior" non-Anglo- Saxon, Roman Catholic immigrants who are prone to drunkenness and who are drawn to socialism. He wanted to curb immigration drastically, a position fueled by his anti- Roman Catholic prejudice.

Strong's most fundamental outlook was a combination of racism, nationalism, and religion. He affirmed passionately the superiority of the "Anglo-Saxon race." And within that race, there is a "natural" elite, namely the Americans. One of the ways in which Strong argues for American superiority is by claiming that there is a "higher" development in physical size among Americans: they are bigger than other people. It is the United States' mission and divine calling to spread a "pure, spiritual Christianity."

Like many clergy of his day, Josiah Strong enthusiastically endorsed the Spanish- American War in 1898. As the United States became a superpower, he did not flinch from his fervent desire for an American Empire. In keeping with this position, he advocated outright annexation of the Philippines during the national rebellion that followed the victory of the United States in the Spanish-American War.

In Josiah Strong, we see the prototype of a theological appropriation of Darwinism and Social Darwinism's preoccupation with the idea of "the survival of the fittest" taking a racist and imperialistic turn. Strong is often described as one of the proponents of the Social Gospel which tried to apply the gospel to the social problems of the day. While I do not want to think that his racism and xenophobia are representative of the best of the Social Gospel movement, they are a reminder the movement had an often forgotten, neglected dark side. It is to the more laudatory aspects of the Social Gospel movement that we now turn.

THE SOCIAL GOSPEL

In contrast to the Social Darwinists, a larger contingent within the Social Gospel movement was quite disillusioned with suffering caused by "laissez-faire" economic policies of the late nineteenth, early twentieth centuries. Although most rejected socialism, all of the adherents of this trajectory in the Social Gospel movement advocated a strong role for government in economic life, supporting such Progressive-New Deal era reforms as minimum working conditions, child labor laws, a minimum wage, limit on hours worked, and the right of workers to unionize and to bargain collectively. Thus, all adherents of this side of the Social Gospel advocated reforms that, by European standards, were "social democratic" or "democratic socialist" and resembled the Christian Socialists of Europe, particularly those of Great Britain, some of whom we shall deal with at greater length in the next section. A few became more radical socialists. Two of the prototypical examples of those who rejected Social Darwinism within the Social Gospel movement were Washington Gladden (1836–1918) and Walter Rauschenbush (1861–1918).

Washington Gladden entered the ordained ministry of the Congregational Church, serving churches in Brooklyn, New York and North Adams, Massachusetts. Between 1871 and 1875, he edited the enormously influential weekly *The Independent* in New York, with a readership of 1 million. The influence of Gladden as a "public intellectual" cannot be minimized.

He served a pastorate in Springfield, Massachusetts 1875–1882, at which time he took on the pastorate of First Congregational Church in Columbus, Ohio, where he would remain for the rest of his life. Gladden was elected to the City Council. Throughout his career he attempted to bring the gospel to bear on socio-politico-economic issues involving labor, poverty, and race. He advocated the kind of Progressive-New Deal era policies described above. Gladden got personally involved in two railroad strikes in Cleveland, Ohio, unequivocally defending the strikers' rights to unionization and arbitration.

Gladden's most famous works include *Applied Christianity* (1876), *Tools and Man* (1893), *The Labor Question* (1893), *Who Wrote the Bible?*, and *Social Salvation* (1902).

Like the other Liberal Protestants of his era whom we have studied, Gladden espoused a form of evolutionary theism and believed in an evolution-inspired progress— provided the principles of the Social Gospel are applied and the concomitant social reforms enacted. Not only did he embrace a form of evolutionary theism, he was enthusiastic about the use of the tools of historical research toward the Bible. The following writing about the Bible provides an example of this aspect of Gladden's thought, which he fused so powerfully with the Social Gospel:

It is not infallible scientifically. It is also idle to try to force the narrative into an exact correspondence with geological science. It is a hymn of creation, wonderfully beautiful and pure; the central truths of monotheistic religion and of modern science are involved in it: But it is not intended to give us the scientific history of Creation, and the attempt to make it bear this construction is highly injudicious." (Gladden 1894, 351)

The name of Walter Rauschenbusch is virtually synonymous with the Social Gospel. He was born in Rochester, New York, his German Baptist father was a professor at Rochester Theological Seminary. With a conservative German Baptist upbringing, he was educated both in Germany and the United States. He was an alumnus of both the University of Rochester and Rochester Seminary (Livingston 1997, 290).

The experiences most formative for Rauschenbusch's career and for his theology of the Social Gospel were provided by eleven years as pastor of Second German Baptist Church in New York City. The Church was located on the edge of Hell's Kitchen, one of New York's most notorious slums. He was confronted daily with the worst of human suffering and despair as well as what he felt was the inadequacy of the inherited tradition to deal with such conditions (Livingston 1997, 291).

Rauschenbusch became a professor in the German faculty of Rochester Theological Seminary in 1897. In 1902, he was given a regular appointment as Professor of Church History. He would hold this position until his death in 1918. Rauschenbusch's major works are *Christianity and the Social Crisis* (1907), *Christianizing the Social Order* (1912), and *A Theology for the Social Gospel* (1917) (Livingston 1997, 291).

Theologically, Rauschenbusch drew freely from numerous sources: "Ritschlian liberalism," named after the German theologian Albrecht Ritschl (1822–1889), which was not concerned with dogma or metaphysics but with the historical Jesus and the historical movement of his followers, especially in relation to ethical problems, which were deemed central for modern Protestantism; evolutionary theism; and the Progressive socio-politico-economic ethos of his American context (Livingston 1997, 291).

The heart of Rauschenbusch's understanding of the Social Gospel revolves around the notion of the "social" or of the "solidaristic." For him the central message of Jesus is about the Kingdom of God, which is not to be identified with individual salvation or in purely spiritual terms but in thoroughly social and historical terms. The Kingdom of God was the most basic *motif* of the Old Testament as well as the message of Jesus and if that was so, the traditional scheme of sin and salvation needs to be rethought in social and solidaristic terms (Livingston 1997, 292). Seeing a weakness in liberal theology's proclivity to minimize the importance of sin, he disagreed with those of its advocates who wanted to do away with that doctrine (Livingston 1997, 292). Characteristically, Rauschenbusch thought the problem would be resolved by a new effort to reconceive sin in social terms (Livingston 1997, 292).

Thus, while as a good Liberal Protestant he rejected the notion of a literal Adam and Eve, the social and solidaristic understanding of sin led him to claim that sin was nevertheless in part biological inheritance and even more importantly, socialization. Thus, in a very real way, the "sins of the fathers (and mothers) are visited unto the seventh generation," to allude to the sensibilities of Hebrew Scripture (which Christians historically have called the Old Testament). The meaning of sin for Rauschenbusch is egoism or selfishness, that is to say having an "unsocial" or "anti-social mind" (Rauschenbusch 1917, 50), which to him was exemplified by the concentration of wealth and power in the hands of the few and in leaving one's peasant laborers intimidated, without dignity, and no claims to the land (Rauschenbusch 1917, 50).

An indispensable part of the rethinking of sin in social terms is rethinking the idea of God. Here, Rauschenbusch, like Abbott, wants to shed hierarchical and monarchical models of God. In fact, he is adamant that we need to " . . . democratize our conception of God" (Rauschenbusch 1917, 49). He asserts that "our universe is not a despotic Monarchy, with God above the starry canopy and ourselves down here; it is a spiritual commonwealth with God in the midst of us" (Rauschenbusch 1917, 49). In democratizing the concept of God, Rauschenbusch is appropriating evolutionary theism's immanent God who is not in absolute control of events.

The Baptist theologian thought of redemption, as well as sin, in social terms. The "composite personalities" and "superpersonal forces" of corporations and the state need to be converted from their collective egoism, the Law of Mammon, the unbridled acquisition of wealth, the monopoly of wealth and power in favor of the Law of Christ (Rauschenbusch 1917, 117). In the case of governments and ruling elites, a similar step is taken when they are truly "democratized," with genuine checks and balances on the powers of corporations, communities, and individuals and when the opportunity to participate in decisions that affect one's life is maximized (Rauschenbusch, 1917, 117). This also meant the enactment of Progressive-New Deal type legislation although Rauschenbusch was more willing than most of his fellow Social Gospelers to take on the label and to advocate "socialistic" measures.

My treatment of Rauschenbusch and the Social Gospel Albrecht Ritschl (1822–1889) leads me mention of Albrecht Ritschl, the father of the Social Gospel who had an enormous impact on American theology as well as his homeland, most evident in such theologians as Wilhelm Herrmann and Adolf von Harnack, Ritschl was a historical and systematic theologian whose most famous work was *The Christian Doctrine of Justification and Reconciliation* (three volumes, 1870–1874). In terms of the theme of this book, Ritschl's importance can be seen in three key emphases in his thought. One of these is placing the Kingdom of God at the center of his theology. The second is evident in his focus on religion being primarily a practical and ethical concern, delineated within the framework of Kant's epistemology and ethics. Third, with his appropariation of Kant's dualism, he paves the way for neo-orthodoxy's independence-contrast model of the relationship between science and religion (Livingston 1997, 270–281).

It may seem strange to include representatives of the Social Gospel in a book about the Liberal Protestant embrace of modern science, in this chapter about the theory of evolution. Yet in the work of people like Washington Gladden and Walter Rauschenbusch, the Liberal Protestant embrace of modern science is part and parcel of the Social Gospel. In particular, a form of evolutionary theism is indispensable as a theological underpinning for their respective theological programs as well as the mainstream of the Social Gospel movement.

Perhaps even more important, this mainstream represented an alternative to the Social Darwinist side of the Social Gospel movement. There were, it goes without saying, significant differences between these two trajectories in their use of Darwinism. The one I would like to highlight is the singular emphasis put on *competition* by the Social Darwinists while the mainstream of the Social Gospel movement focused on *cooperation*. For the Social Darwinists, survival of the fittest meant *competition* even within one's own species, "nature red in tooth and claw" even in economic competition, while for the likes of Gladden and Rauschenbusch and their followers survival of the fittest meant *cooperation within* a particular species if that species wanted to survive given the competition *between* species for scarce food supplies. This emphasis on cooperation rather than competition as a prerequisite for the survival of the fittest was typical of those social reformers who appropriated the insights of Darwinism. Perhaps the most consistent and articulate expression of this view is the Russian anarchist Petr Kropotkin's *Mutual Aid*, although it needs to be stated that no causal connection seems to exist between it and the mainstream Social Gospelers we have studied.

LUX MUNDI

Even when Bishop Wilberforce was having his famous debate of 1860 with Thomas Huxley, there was considerable support in the Church of England for Darwin's theory of evolution. For example, Darwin had an active correspondence with his popular and influential priest-theologian friend Charles Kingsley. The latter thought that the theory of evolution provided an opportunity to conceive of God as imminent in creation, ever present, ever working in the most intimate way with her/his creatures, "making them make themselves."

Within thirty years of the publication of *The Origin of Species*, a group of the most important and influential theologians of the Church of England published a selection of essays edited by Bishop Charles Gore with the title *Lux Mundi* (*Light of the World*). The essays presupposed the truth of evolutionary theory and some very explicitly sought to explain, even recast the faith in terms of Darwinian theory.

The introductory essay in *Lux Mundi* by H. S. Holland on "Faith" provides an excellent example. For Holland, science is a friend and not an enemy of faith (Holland 1909, 26). For faith, scientific knowledge, in fact all secular knowledge can only be a gain (Holland 1909, 26–27). Even as faith can affirm and take into itself all knowledge, so it can help us recognize the ever changing, tentative nature of all knowledge (Holland, 1909, 27).

In his essay in *Lux Mundi*, "The Incarnation in Relation to Development," J. R. Illingworth, for example, maintains that while modern science may show us that everything is energy, it is theology that demonstrates that all energy comes form God (Illingworth 1909, 137). Another major contribution of then contemporary science was to show that the world is not a machine but " . . . an organism, a system in which, while the parts contribute to the growth of the whole, the whole also reacts upon the development of the parts" (Illingworth 1909, 139). This world of order and beauty points to a teleology within the process of natural selection that ultimately points to the Eternal Reason we call God (Illingworth 1909, 139).

Even more importantly, echoing the Liberal Protestants we have studied so far, Illingworth claims that modern science illustrates " . . . [God's] indwelling presence in the things of His creation" (Illingworth 1909, 139). In this regard, modern science and Darwinism in particular rescued modern theology from the distant, absent God of deism (Illingworth 1909, 139-140). In his estimation, the facts are with evolution and they are not a threat to the Christian faith (Illingworth 1909, 141–143).

In "The Incarnation as the Basis of Dogma," R. C. Moberly takes on the issue of whether or not church dogma is based on evidence, experience, and reason or is blind, unquestioning assent. To him, the latter has nothing to do with faith nor with "the fullest exercise of intellect" required for theological reflection (Moberly 1909, 160–161). Moreover, science has its own set of dogmas.

Moberly claims that church dogma is based on evidence, experience, and reason. Although there is unity to all knowledge, there is a distinctiveness to both scientific and religious knowledge. The latter speaks to the whole person, to the heart, and motivates moral action in a way that science cannot (Moberly 1909, 160–167).

Writing in a chapter entitled, "The Holy Spirit and Inspiration," the editor, Bishop Charles Gore, sounds a familiar refrain of Liberal Protestantism, drawing on both modern science and the tradition of the Church Fathers in the early centuries of the history of Christianity: "Nature is one great body, and there is breath in the body; but this breath is not self-originated life, it is the influence of the Divine Presence" (Gore 1909, 232). Gore, true to the spirit of his times and to that of most of the Western tradition, remains anthropocentric, human centered, highlighting the special role of humans in creation.

The prototype of the Anglican embrace of modern science and Darwinian evolution in particular is Aubrey Moore (1848–1890). Typifying this attitude, in "The Christian Doctrine of God" in *Lux Mundi*, Moore makes his famous statement: "Science had pushed the deist's God farther and farther away, and at the moment when it seemed as if He would be thrust out, Darwinism appeared, and, under the disguise of a foe, did the work of a friend" (Moore 1909, 73). He maintains that Darwinism has given both philosophy and religion an inestimable gift in confronting people with two choices. He elaborates:

Either God is everywhere present in nature, or He is nowhere. He cannot be here and not there... We must frankly return to the Christian view of direct Divine agency, the immanence of Divine power in nature from end to end, the belief in a God in Whom not only we, but all things have their being, or we must banish Him altogether. It seems as if, in the providence of God, the mission of modern science was to bring home to our unmetaphysical ways of thinking the great truth of the Divine immanence in nature, which is not less essential to the Christian idea of God than to a philosophical view of nature. (Moore 1909, 74)

Moore contends that the notion of divine immanence had attained such a popularity in his era that the Christian faith was threatened as much by pantheism, the idea that God is the totality of things or, in more contemporary renditions, the web of life itself, as it was by deism. It is interesting to note that he mentions John Fiske as an example of such pantheism. For Moore, upholding the divine immanence in all things yet not identifying God and the world is one the major challenges theology faces (Moore 1909, 74–75).

Moore was also highly critical of the position that espoused God's special acts of creation, exemplified in the creation of new species, as seen in a literal reading of the Genesis account and in the traditional notion of the fixity species. Not only did such a view defy scientific evidence, it also reinforced a sense of God's absence, " ... that *a theory of occasional intervention implies as its correlative a theory of ordinary absence* ... " (Livingston 1997, 258). He also rejected the argument from design, finding it refuted by the destruction evident in nature. He believed that Christians need to trust in God's goodness just as scientists need to trust in the rationality of the universe (Livingston 1997, 258). Moore, in a way that anticipates contemporary theological efforts at overcoming anthropocentrism, rejects traditional views of humanity's place in the universe, that is to say its superiority to other species. To see humanity as coming from the lowest origins, from the dust, is not belittling or degrading to humanity but edifying. And, I might add, that it is profoundly Christian (Livingston 1997, 258–259).

In discussing Anglican theologians, I would be remiss not to mention the Episcopal Bishop Phillips Brooks (1835–1893), one of the great preachers in nineteenth century America, perhaps most famous for being the author of the Christmas carol "O Little Town of Bethlehem." In contrast to devotees of the High Church, who advocated the retrieval of more formal ritual (yet saw evolution and development in both the doctrine and liturgy of the church) and the Low Church, Evangelical who emphasized the primacy of scripture although not necessarily in a literalist way, Brooks was a spokesperson for the Broad Church movement. Not only did the Broad Church seek a middle way, it sought to do so by using contemporary knowledge, especially as derived from modern science to do so.

The appropriation of Darwin and the theory of evolution in these Anglican theologians fits nicely with the incarnational-sacramental ethos of Anglicanism. The incarnational ethos of Anglicanism is based on the famous passage in the Gospel of John (3:16) that God so loved the world that God became one of us in Jesus Christ, sharing our lives and our deaths. In similar fashion, God uses the "stuff" of the earth and, empowering humans to cooperate, as in making the bread and the wine from elements of creation, material things to disclose the divine self and to draw us to live in that divine self. Moreover, in Anglicanism, there is a stress on worshipping God "in the beauty of holiness." That is to say, every human endeavor is to be used for the greater glory of God, there being nothing outside of the divine purview. Thus, Anglicanism is a radically world affirming tradition. In appropriating Darwin and the theory of evolution, the illustrious theologians who contributed to *Lux Mundi* were intentionally and self-consciously following this most fundamental sensibility.

A persistent issue that rises to the forefront in the late nineteenth century liberal embrace of modern science and the theory of evolution—and one that is very much in the forefront of the contemporary religion and science dialogue—is the issue of "supernaturalism" vs. "naturalism." The Latin preposition "super" literally means "over" and "beyond." Historically, "supernaturalism" has meant "over" or "beyond" the physical world, as in the supernatural, personal God, traditionally, over and beyond the world, intervening in the world from "the outside." "Naturalism" means that there is no world beyond this one; if there is a God, she/he is in the world or nowhere. Certain forms of naturalism are materialistic and reductionistic; God disappears from the picture. We saw this in the last chapter, although we did not necessarily call it naturalism. There are a variety of other forms of naturalism which see the world as dynamic and creative rather than as a machine, with room for some concept of God, whether as the laws of nature or something else, operating *within* the world of nature.

Figure 3.2 Bishop Phillips Brooks, on *The Phillips Brooks Calendar for 1898* (Courtesy Library of Congress)

Among the Protestant Liberals we looked at in this chapter, clearly there are some important figures who unequivocally take the side of naturalism—Lyman Abbott is one glaring example. There are those, on the other hand, like Charles Gore, who tried valiantly to synthesize a naturalistic view of nature with a supernaturalistic understanding of God. Whether or not this is possible or whether or not science makes some sort of naturalistic understanding of God necessary is an issue that still haunts theology today—and is one that we shall encounter again in a subsequent chapter.

We have looked in this chapter at Liberal Protestantism's radical embrace of the theory of evolution. This radical embrace was part and parcel of an ethos that in part reflected, in part helped create the optimism and confidence in progress of the late nineteenth century. This optimism and confidence would be shattered in Europe by the assassination of the Archduke Franz Ferdinand in Sarajevo on June 28, 1914, which led to the outbreak of World War I, while in America they would be shaken by the Great Depression. Before we look at the impact of these disillusioning cataclysmic events on Liberal Protestantism's relationship to modern science, we shall turn to the convergence of religion and science in some Liberal Protestant theologians of the late nineteenth, early twentieth centuries.

I want to conclude this chapter with perhaps what is the most powerful image of Liberal Protestantism's radical and enthusiastic embrace of the Darwinian theory of evolution. Let me begin with a question: Do you know where Charles Darwin is buried? If you thought or said, "Westminster Abbey!" you are correct. Who makes the decision as to who gets buried in Westminster Abbey? It is the Church of England. On account of his singular contributions to human knowledge, the Church of England conferred its highest honor on Charles Darwin!

Chapter 4

—

Science, Liberal Protestantism, and the Twentieth Century: 1900–1960

THE CONVERGENCE OF SCIENCE AND RELIGION

My primary example of the convergence of science and religion in the late nineteenth, early twentieth centuries, focuses on the liberal theologians who came to be called "the Chicago School," not to be confused with the more "conservative" thinkers associated with "Chicago Schools" in the Economics and Government Departments. The University of Chicago was established by the American Baptist Education Society and endowed by John D. Rockefeller. It opened its doors in 1892. William Rainey Harper was its first president.

Harper was a Hebrew scholar who wrote several volumes about the Old Testament. As president, he had considerable success in attracting some of the greatest scholars to the fledgling university, especially to the Divinity School. Just as Chicago came to symbolize the changing fulcrum of the country, so did the University of Chicago symbolize the beginnings of the shift (or at least of diffusion) of the intellectual center of the country away from Harvard and the Northeast.

The theologians Harper gathered at the Divinity School, "the Chicago School," developed a distinctive way of doing theology, the socio-historical method. According to this method, the Christian tradition was not something permanently etched in stone, eternal and unchanging, but was an ever-changing historical movement that interpreted and reinterpreted itself as it responded creatively to the ever-changing challenges of the present. In utilizing the socio-historical method, the members of the Chicago School appropriated the insights of the natural as well as social sciences. Among the illustrious members of the school were George

Figure 4.1 William Rainey Harper (Courtesy
Library of Congress)

Burman Foster, Shailer Mathews, Edward Scribner Ames, Gerald Birney
Smith, Shirley Jackson Case, and others too numerous for adequate treatment in this volume.

The first Chicago School theologian I shall consider is George Burman
Foster. He was born in West Virginia in 1858. Foster, ordained to the Baptist
ministry in 1879, attended West Virginia University, receiving the B.A.
and A.M. degrees respectively in 1883 and 1884. He got married in 1884
and enrolled at Rochester Theological Seminary, from which he graduated
in 1887. The academic year 1891–1892 that Foster spent in Gottingen and
Berlin provided the context for the development of his early Ritschtlianism.
He taught at McMaster University in Toronto upon his return until 1895, at
which time he moved to the University of Chicago Divinity School, where
he taught until he died of an infection in 1918.

Even before the publication of his first book, Foster was a controversial
figure. To the relief of Harper, who was under considerable pressure from
Baptist clergy to get rid of Foster, the latter requested reassignment from
the Divinity School to the university as professor of philosophy of religion
in the Department of Comparative Religion in 1905 (he was succeeded

in his chair at the Divinity School by Shailer Mathews, who would be no stranger to controversy himself). The two books the controversial theologian published during his lifetime, *The Finality of the Christian Religion* (1906) and *The Function of Religion in Man's Struggle for Existence* (1909) created even more of a stir (his serving a Unitarian Church in Wisconsin on weekends did not help matters), culminating in his being put on trial by and expelled from the Baptist Ministers' Conference (*Christianity in Its Modern Expression*, 1920, and *Friedrich Nietzsche*, 1931, both appeared posthumously). He did remain an ordained Baptist minister and retained his position in the faculty at the University of Chicago as well as his membership at the Hyde Park Baptist Church. Throughout the controversies, he defended the integrity of his position as in keeping with the spirit of free inquiry and the independence of the Baptist tradition.[1]

Foster embraced the modern world, particularly modern science, its methods and findings. As was typical of the adherents of religious liberalism, he wrestled with the question of how one could be a modern person and a Christian at the same time. Central to his consistent way of dealing with the question throughout his career was the rejection of what he considered the "authoritarianism" of "supernaturalism" as well as "naturalism." The "naturalism" he rejected was the mechanistic, deterministic, reductionistic worldview that has characterized much of modern scientific materialism. He did espouse the notion of a dynamic, creative, evolutionary universe that has room for spirit—as we shall see, much like what Daniel Day Williams has called "neo- naturalism."

In his early work, Foster was a Ritschtlian profoundly influenced by Kant. At this point, the person and teachings of Jesus as well as an emphasis on morality were central to his thought. The traditional proofs for God's existence were no longer valid and, besides, were not what Jesus taught. Neither were dogmas, for that matter. Rather, Jesus proclaimed the Kingdom of God, which we are called to co-create as we realize our vocation to become persons. Just as he realized his personhood in his inner communion with God expressed in loved for others, so are we called to realize ours, in our own way, through our own "God-consciousness" lived out in loving others (Foster 1906).

---

[1]For biographical information, see Alan Gragg, *George Burman Foster: Religious Humanist* (Danville, VA: Perspectives in Religious Studies, 1978); Creighton Peden, *The Chicago School: Voices in Liberal Religious Thought* (Bristol, IN: Wyndham Hall Press, 1987), pp. 24–43; Edgar A. Towne, "Introduction to Foster," in W. Creighton Peden and Jerome A. Stone, eds., *The Chicago School of Theology—Pioneers in Religious Inquiry, Vol. I, The Early Chicago School: G. B. Foster, E. S. Ames, S. Mathews, G. B. Smith, S. J. Case* (Lewiston/Queenston/Lampeter: The Edwin Mellen Press, 1996), pp. 1–5, which includes an extensive bibliography of works by and about Foster; Gary Dorrien, *The Making of American Liberal Theology: Idealism, Realism, and Modernity, 1900–1950* (Louisville, KY: Westminster John Knox Press, 2003), pp. 151–181.

By the time Foster published *The Role of Religion in Man's Struggle for Existence*, he had taken an empiricist, functionalist turn in his theology. His focus is on the interaction between organisms and their environments out of which personality and the soul emerge. In the vocation to become persons, religion and "the gods," human constructs, are indispensable. While at times sounding like a critical realist, there is such strong element of subjective interpretation in Foster that one wonders if there is any independent reality to God and "the gods" aside from their functional utility and the human proclivity to create them in our own image. God is a symbol that designates the ideal achieving capacities of the universe (Foster 1909).

Foster's theological move has often been described as one from Ritschtlianism to "religious humanism." To religious humanists, what matters in religion is how it aids humanity and human fulfillment. There is little if any concern with God and other theological doctrines except as social constructs that either aid or impede human self- realization. In contrast to this assessment of Foster, Edgar Towne has maintained that the Chicago theologian's move was " ... beyond the dualism of super-naturalism/naturalism to a theology of spontaneity (childlikeness), freedom (moral and intellectual integrity), and personality (Christlikeness)" (Towne 1996, 3–4).

It needs to be noted that Foster was not only a tortured soul who anguished over how one could be a modern human being and a Christian at the same time but also one who experienced tragedy profoundly throughout his life: his mother died when he was five years old; he was left with his paternal grandparents when his father fought in the Civil War; and all five of his children, two suffering from mental illness, died in the prime of life. During the controversies with the Chicago clergy he received only one letter of support from his fellow clergy—from a rabbi! It would not be exaggerating to say that Foster was literally hounded to death by his theological opponents.

It is not widely known that Foster's best friend was Clarence Darrow (1857–1938). Darrow is most famous for his role in the Scopes Trial, popularized by the play and motion picture, *Inherit the Wind.* As such, he has become a symbol for the supposed "eternal warfare between science and religion."

Darrow was certainly also justly famous for his defense of dissenters, of union organizers, and of unpopular causes: Eugene V. Debs, the future Socialist presidential candidate who organized the Pullman strike; William Haywood, president of the Industrial Workers of the World (I.W.W.), who was tried for assassinating the governor of Idaho; and Leopold and Loeb. His fame was not without ambiguity: he was tried once for attempting to bribe a juror.

Darrow and Foster had a series of debates on free will and on whether life is worth living. Their erudite yet warm and witty exchanges provide masterful examples of the beauty of the English language. Darrow gave the eulogy at Foster's memorial service. The friendship between Clarence Darrow and George Burman Foster provides a powerful symbol for the relationship between science and religion—one that is quite different from the usual image of Darrow and the supposed "eternal warfare between science and religion." While one could maintain that Foster's theology reflects the result of the felt conflict between religion and science, I see him as anguished in trying to sort out the relationship between the two while doing justice to each. I have given him such a lengthy treatment because he is the prototype of the Liberal Protestant commitment to the spirit of unfettered intellectual inquiry and letting the chips fall where they may.

Alluding to the Divinity School of the University of Chicago and its notable array of theologians who were engaged in the use of the socio-historical method, John Cobb has written that " ... the man who best typifies the Divinity School and most influenced its development through the first three decades of this century was Shailer Mathews" (Cobb 1982, 22–23). Mathews (1863–1941) went to the University of Chicago Divinity School in 1894, becoming Dean in 1908, a position that he held along with those of Chair of the Department of Theology and Ethics and Professor of Historical Theology, until his retirement in 1933. A prolific scholar who, with William Rainey Harper, the first president of the University of Chicago, sought to "democratize biblical scholarship" by making it accessible to lay people, he was active in his own Northern Baptist Convention, serving as its president, and in ecumenical organizations. He was also active in the Progressive wing of the Republican Party (long before that species became extinct!) and supported reform candidates in Chicago's mayoral politics. He took a leave of absence from the Divinity School to sell war bonds. Few American theologians have been so influential in public life.

It is helpful to treat Mathews' thought as a tripod the three legs of which are the socio-historical method, modernism, and the social gospel. The socio-historical method attempts to situate concrete events and peoples within the larger context of social movements and civilizations that respond creatively or destructively to felt needs and challenges. Through his use of the socio-historical method, Mathews wound up viewing religious traditions as continuous movements that appropriate and reappropriate the past as they respond creatively to contemporary social needs. In keeping with this notion, for him Christianity is a historic group movement rather than a set of eternal truths or doctrines. Doctrines and theologies are particular responses to particular historical contexts, reflections of as well as responses to socio-political-economic changes.

The best examples of Shailer Mathews' use of the socio-historical method are his *The Atonement and the Social Process* (Mathews 1930) and *The Growth of the Idea of God* (Mathews 1931). A study of great breadth, the former examines how various theories of the atonement were responses of the Christian community to the particular socio-politico-economic problems of specific eras. Similarly, the latter explores how particular concepts of God are profoundly shaped by particular experiences of political sovereignty, functioning, in effect, as what Mathews called a form of "transcendental-ized politics" (Mathews 1930, 92, 94).

Just as the Christian tradition has always appropriated and reappropri-ated the past as it responded creatively to the needs of particular historical periods, so does it need to appropriate the inherited tradition even as it responds creatively to the needs of the present. Responding adequately to the needs of the contemporary world, according to Mathews, entails the unequivocal embrace of modernism.

The modernism that Mathews embraces is thoroughly functional and pragmatic. He claims that theology itself is functional, asserting that "a theological pattern of unchanging content has never existed" (Mathews 1924, 72). Moreover, "when a pattern no longer expresses a religious value or serves as the symbol of a group attitude, it should be and has been abandoned" (Mathews 1924, 72) in order to respond creatively to the needs of the present. Mathews defines modernism as " ... the use of the methods of modern science to find, state, and use the permanent and central values of inherited orthodoxy in meeting the needs of a modern world" to the point that "the needs themselves point the way to formulas" (Mathews 1924, 23). Modernism also emphatically rejects any form of dogmatism and authoritarianism, allowing freedom of inquiry to lead wherever it may (Mathews 1924, 23–24).

Mathews thought that it was by Christianity's embrace of modernism that the Gospel would become intelligible to contemporary women and men. In the very endeavor, it would also be evangelical, in his estimation (Mathews 1924, 34–35)!

Mathews selects two paradigms for the reconstruction of Christian life in a way that is congruent with contemporary experience. The first of these involves the reformulation of the concept of God in ways that are consis-tent with democratic experience and ideals (Mathews 1924, 106–108). The second is that of the dynamic interrelatedness of and interaction between organisms and their environments as discovered by modern science.

If the human personality, which issues from the interaction between organisms and their environments, is the most complex form of life, as far as we know, to have emerged in a cosmic creative evolutionary process that sustains life, then God is identical with the "personality producing elements" of the universe; the point of the Christian life, of which doctrines are particular expressions, is adjustment to those personality producing forces of the cosmos (Mathews 1930, 185–186; 1931, 208–209; 1940, 19– 52). Sin, with its devastating personal, socio-politico-economic-cultural, and cosmic consequences, is maladjustment to these elements. Consistent with his affirmations about God and sin, Mathews sees Jesus as the supreme instance of the cooperation between the personality producing elements of the cosmos and human beings (Mathews 1930, 191). Hence, salvation is adjustment to the personality producing elements revealed in Jesus that work persuasively toward moral transformation in individual and social existence.

Anticipating some of the trajectories of postmodernism, Mathews takes a non-essentialist approach to Christianity. That is to say, Christianity (or any tradition, religious or secular, for that matter) as a historic movement has no "essence," no "kernel" above or beyond the vicissitudes of history, no "core" below the time-bound layers of tradition. He is nevertheless concerned with the maintenance of the "permanent values" of Christianity (Mathews 1936, 49) and with "the basal Christian attitudes and convictions" that are "the permanent element in our religion" (Mathews 1924, 82), which he claims can be found in "active loyalty to Christ and his message that God is fatherly and that men therefore ought to be and can be brotherly" (Mathews 1924, 82). Extending these "permanent values" into every facet of life is inseparable from Mathews' understanding of modernism and is at the very heart of his understanding of the social gospel.

The Chicago theologian affirms a relational rather than an atomistic view of the human self. Thus, salvation is social as well as individual. Consequently, this means that society needs to be so organized as to nurture the growth of the human personality. Mathews bases these claims on what seems to be a link between his love of democracy and his understanding of Jesus' teachings and identification with the poor. "The democratization of privilege," one of Mathews' favorite phrases, is a metaphor for his support for the social welfare reforms advocated by the Progressive movement and enacted during the New Deal, social welfare reforms he deemed vital for fostering the development of the human personality.

Mathews' development of the concept of God as the personality-producing elements of the universe is within the context of his embrace of modern science, the theory of evolution in particular. However, in embracing the theory of evolution, the emergence of complex forms of life from the simple, the Chicago theologian is also attempting to situate humans in the non-human natural world even as he attempts to preserve the distinctiveness of the human personality. He is adamant about the inadequacy of seeing human and non-human creatures alike in mechanistic terms (Mathews 1930, 185–186). He uses the term "organism" to refer to living things in contrast to mechanistic reductionism, one form of which is to reduce living things to the manner in which chemicals come together and interact in various bodies (Mathews 1930, 185–186). And with the term "personality" he highlights the distinctiveness of the human, in contrast to non-personal organic existence (Mathews, 1930, 186, 1940, 29–32). Remarkable for his time in situating humans in the non-human natural world even as he sought to safeguard the distinctiveness of the human, Mathews' efforts to highlight human personality seem anthropocentric, especially given his unfortunate rhetoric about the "sub-human" and its connotation of non-human inferiority (Muray 1996, 119–125).

For Gerald Birney Smith, as for Foster and Mathews and the other great theologians at the University of Chicago, religious traditions are continuous social movements that appropriate and reappropriate the past as they respond creatively to contemporary social needs and challenges. The greatest contemporary challenge, in his view, is the chasm between authoritarian ecclesiastical control of religious, political, and intellectual life, the dominance of deductive reasoning of previous ages, and the prevalence of inductive reasoning, of scientific reason, and the increasing spread of democratic procedures in the modern age. His own constructive theology is a response to this challenge in its radical affirmation of the spirit of inquiry, inductive reasoning, and democratic processes, most evident in his endeavor to "democratize" religion and to develop the notion of a "democratic God."

In his attempt to "democratize" religion, especially Christianity, he advocates democratic structures in ecclesiastical organization, non-hierarchical, participatory modes of ordained and lay ministry, all hinging on a "democratic" concept of God. It goes without saying that there is a profound and insoluble link between his notions of democracy and God and his endeavor to "democratize" religion, suggesting the contours of a consistent if embryonic systematic theology, all parts of a response to modern science.

To gain a better understanding of what he is attempting to do, we need to examine Smith's characterization of "democracy." The Chicago theologian claims that the first ideal of democracy is "the right of revolution against autocracy" (Smith 1919, Vol. 53, 7). He writes that "the most important thing about democracy is denial of the right of the class system to persist" (Smith 1919, Vol. 53, 7). Smith is here reflecting on the bloody historical development of democracy (Smith 1919, Vol. 53, 7–13). In this regard, he writes that "democracy means that the people claim for themselves the right to determine what is just" in contrast to the most basic principle of autocracy, " ... the right of a superior to determine for his inferiors what they should do" (Smith 1919, Vol. 53, 8). For an inferior to question this is to challenge the status quo, to rebel against properly "constituted authority" and the "very nature of things" (Smith 1919, Vol. 53, 8).

Smith describes the interplay of legitimizing symbolic and mythic structures in this complex historical development in the following manner: "If now, as has generally been the case, the autocratic order is believed by those who adhere to it to be divinely established, the democratic revolution looks like a defiance of God's laws" (Smith 1919, Vol. 53, 7). In such cases, democratic ideals have expressed themselves through a radical criticism of the religion that supports and legitimates "l'ancien regime," "the old regime." Anticipating his constructive endeavors, Smith points out that there is nothing intrinsically irreligious about democratic aspirations. He asserts that "indeed those contending for popular rights often invoke a divine sanction for their attempts; but even so, it involves a different kind of religion from that of the established order" (Smith 1919, Vol. 53, 8).

The right of revolution depends on the fundamental principle of equal rights for all (Smith 1919 Vol. 53, 8). Smith affirms this principle eloquently in maintaining that "the moral defense of revolution consists in establishing the doctrine that there is inherent in human nature a dignity which entitles all men to equal rights" (Smith 1919 Vol. 53, 10). Indeed, he sees the progressive removal of the vestiges of autocracy, such as the abolition of slavery, of religious tests as a basis of citizenship, and the extension of the franchise to women as hallmarks of enduring democracies (Smith 1919, Vol. 53, 10).

Another of the central ideals of democracy is "the responsibility of citizens for good government" (Smith 1919, Vol. 53, 10). Smith states in this regard that "the full meaning of democracy is realized only when the citizens become conscious of themselves, not as claimants for special benefits, but as responsible partners in the conduct of a great enterprise for the common good" (Smith 1919, Vol. 53, 12).

For Smith, the final ideal of democracy is democratic control of "special ability." What he has in mind specifically is business and industry in which he sees a fundamentally anti-democratic hierarchicalism, with employers and managers often acting as the "new aristocracy." Briefly, he suggests that we search for institutional avenues in which both the "special ability" of entrepreneurs and employers is preserved and the "special ability" of employees democratically expressed (Smith 1919, Vol. 53, 13).

The closest Smith comes to a definition of democracy is the following:

Democracy is, in essence, the assertion of the right and ability of men to determine for themselves what they want and to control the officials who administer the laws designed to secure the desired ends. This assertion of fundamental human rights has at the basis of any democratic movement. (Smith n.d., 411)

The fundamental premise is that humans have both the right and the ability to exercise freedom, that human welfare is the final criterion of judgment. This is in sharp contrast to the basic premise of much of medieval and Reformation theology that it is from the decree of God, from divine ordinance that everything else must be deduced, and that human nature is depraved or, at the very least, fundamentally tainted by original sin (Smith 1919, Vol. 53, 411).

Implicit in this discussion is the chasm Smith sees between premodern deductive modes of thought and modern inductive, scientific reasoning. If in the former certain things are assumed to be true, are beyond question, and from which all other truths follow, in the latter there is nothing that is beyond question, the search for truth to be followed regardless of wherever it may lead and the consequences to be faced. The tentativeness and openness of this method is indispensable for the development of democracy and for the nurturing of the kind of citizenry that can sustain such a development.

It is in the context of the characterization discussed above that Smith's sense of urgency about the "democratization" of religion needs to be seen. And if this is the greatest challenge facing religious traditions and communities, such a "democratization" of religion cannot exclude our notions of God.

In developing the notion of a "democratic God," Smith begins by claiming that "if the church is to exert its rightful influence in a democratic age, it must enable men to worship God in such a way as to give sanctity to the great ideals of democracy" (Smith 1919, Vol. 53, 632). However, it is profoundly problematic that while "these ideals are concerned with very practical problems of human need and injustice, and with the possibilities of a richer life here and now," the theology of the creeds and to considerable extent the content of rituals reflect a concept of God that draws upon analogies to autocratic regimes (Smith 1919, Vol. 53, 632). These analogies to autocracy are, moreover, closely tied to the veneration of a past that is considered normative (Smith 1919, Vol. 53, 632). As a result, God seems remote and unreal in light of the lived experiences of contemporary women and men (Smith Vol. 53, 632).

Smith points out that analogies to autocracy emphasize the transcendence of God. Conversely, democratic analogies need to emphasize the immanence of God (Smith 1919, Vol. 53, 632). This entails, as the Chicago theologian claims is typically done, much more than merely picturing God in the world rather than above it for this is still the image of a king who has retained all the royal prerogatives unchanged. In a democracy, the monarchy and its prerogatives and attributes either change, as in England, or it disappears, as in France (Smith 1919, Vol. 53, 632). An image

Smith uses to explain his sense of the interconnection between the immanence of God and democratic analogies is the sharp line of demarcation
between king and subject in an autocracy: they belong to different worlds
much in the manner that God and humans belong to different worlds
(Smith 1919, Vol. 53, 633). In contrast, there is no such sharp distinction in a
democracy since, at least ideally, "the ruling power is integrally one with
the citizens" (Smith 1919, Vol. 53, 633). Consequently, an emphasis on the
immanence of God is analogous to the notion that "the ruling power" that
is God is integrally one with the citizens of creation (Smith 1919, Vol. 53,
633).

Although Smith finds the idea of a "finite" God "unhappy" because it is
religiously unsatisfying (Smith 1919, Vol. 53, 633), I would characterize his
understanding of God's power as finite and limited. I characterize his
concept of God in this manner because of his rejection of absolute
sovereignty as fundamental to the character of the divine power and the
virtual analogy of divine power to a "constitutional monarchy" exercising
royal power "in accord with the rules of games," in this instance not above
or beyond but in accord with the laws of the universe (Smith 1919, Vol. 53,
633).

If God is not above or beyond the world but radically immanent in it,
then God is a suffering God who bears the burden of evil in the world—a
relational view of God hinting at panentheism, the notion that God is in the
world and the world is in God (Smith 1919, Vol. 53, 633). It is not straining
the point to claim that Smith makes a similar claim to relationality in his
discussion of "sin" and "salvation." The Chicago theologian defines sin as
" . . . a willing aloofness of men from the welfare of his fellow-men . . . "
(Smith 1919, Vol. 53, 637). "And," Smith asserts, "this aloofness of a man
from welfare is precisely an attempt to withdraw one's self from the real
presence of God, who is working though history for the release of men
from the evils which beset them" (Smith 1919, Vol. 53, 637). Salvation is
the overcoming of this aloofness, the fruits of which are that the individual
is democratized and becomes " . . . a sharer in the life of humanity," with
a sense that "his own welfare can be attained only as the welfare of all
shall be promoted" (Smith 1919, Vol. 53, 633). He summarizes the point by
claiming that "enlistment in a genuinely social life is the very pathway to
God" (Smith 1919, Vol. 53, 637).

Without the certainty of knowing where the divine activity stops and human activity begins, without the assurance of knowing the final outcome
or even that ultimately good will triumph, we are called to be co-workers
with God to be saved from sin (Smith 1919, Vol. 53, 636). Furthermore,
he states that: "and if God is presented as the immanent power working
through the efforts of men to shape history so as to make a better world,

reconciliation with God is at the same time reconciliation with the righteous cause to which religious men are devoted" (Smith 1919, Vol. 53, 636). This God who empowers the overcoming of the sin of aloofness and the salvation of "sociality" does so because she/he is a struggling, suffering God to whom all things make a difference, the supreme instance of "sociality."

A few remarks are in order about pertinent aspects of Smith's understanding of Jesus. Consistent with his emphasis on the immanence of God and rejection of autocratic analogies is that he stresses "the historical Jesus as he lived among men," and deems inadequate the Chalcedonian formulation and its subsequent interpretations which equally emphasize the transcendence of God (Smith 1919, Vol. 53, 633). He maintains that:

It is absolutely indispensable to the efficiency of the saviorhood of Jesus from the point of view of … typical modern interests, that he should enter completely into the perplexing experiences which constitute the religious problem of the modern man. He must be a citizen of this world, rather than an alien from another world. The mere "human nature" of the traditional creeds is incompetent to express the psychological and ethical content which is indispensable if Jesus is to enter as a vital transforming force into the religious experience of modern man. (Smith 1996, 224–225)

In line with this argument Smith highlights the continuity between the sacrifice of soldiers during the course of World War I and the sacrifice of Jesus Christ on the cross—a perceived continuity that, as the Chicago theologian delights in pointing out, no one considered a sacrilege (Smith Vol. 53, 634). It is important to realize that in drawing the comparison between the sacrifice of Jesus and the sacrifice of soldiers in the World War I, he is not glorifying war nor is he a bellicose militarist; rather, in a fashion typical of his constructive theology, he is attempting to reinterpret the cross in terms of the actual, lived experiences of contemporary human beings.

Although Smith had little interest in developing a reinterpretation of the doctrine of the incarnation, the continuity, consistency, and coherence between his respective emphases on the immanence of God and the human, historical Jesus suggest the outline of a systematic theology. And while he is uninterested in addressing systematically such traditional Christological questions as how God is present or incarnate in Jesus of Nazareth, the coherence between Smith's notion of a suffering God who bears the evil in the world and the linkage between Jesus' sacrifice on the cross and the experience of contemporary humans is no less suggestive of a profoundly incarnational theology.

Finally, toward the end of his life, appropriating the insights of science, particularly the theory of evolution, the Chicago theologian's work is increasingly preoccupied with the question of how this "democratic God" is a cosmic reality, a finite reality within nature, of which humans are parts just as the non-human natural world is a part of us. This is the mystical dimension of Smith's democratic religion that is ever present but most prominent at the end of his career. He writes:

The belief in God means that there may be found, not merely within the circle of human society but also in the non-human environment on which we are dependent, a quality of the cosmic process akin to the quality of our own spiritual life. Through communion with this qualitative aspect of the cosmic process human life attains an experience of dignity, and a reinforcement of spiritual power. The quality of this reinforcement can be adequately expressed only by the conception of a Divine Presence in the cosmic order. (Smith 1996, 216)

And it is in this context and profoundly shaped by it that "we must learn to think of God as the immanent coworker always toiling with his children rather than as the sovereign to whom they are subject..." and of "the salvation which God makes possible ... as a process of cooperation with God..." (Smith 1996, 206–207).

There are other important Liberal Protestant theologians/pastors worthy of mention and of these I shall mention two: Harry Emerson Fosdick (1878–1969) and Douglas Clyde Macintosh (1877–1948). Fosdick was the longtime American Baptist Pastor of Riverside Church in New York City, famous for his preaching and his hymnody. He sought to reinterpret the Bible to contemporary women and men in light of modern science. He was in the thick of the modernist-fundamentalist controversy, treating the Bible as a book that shows the progressive revelation of God and the progressive development of humanity's understanding of that revelation.

Macintosh, a Canadian, spent most of his career at Yale University Divinity School, mentoring among others Reinhold and H. Richard Niebuhr. Macintosh attempted to develop an empirical theology, another version of which we shall consider later, that appropriated the insights of and was faithful to modern science. He was probably most famous for being denied U.S. citizenship on account of his pacifism in 1931.

I need to mention briefly that fundamentalism arose as a reaction against the Liberal Protestant embrace of modern science and its concomitant secularity. Although fundamentalism had its antecedents, perhaps especially in the revivalist movement and at Princeton Theological Seminary, we do not encounter the word until the publication of the "The Fundamentals" in 1910. The "five fundamentals" are: (1) the verbal inspiration and inerrancy of scripture; (2) the divinity of Jesus Christ; (3) substitutionary atonement,

Figure 4.2 Harry Emerson Fosdick with the head of the YMCA in 1926 (Courtesy Library of Congress)

i.e., the doctrine of Christ taking on and dying for our sins; (4) the bodily resurrection of Jesus Christ; and (5) the reality of miracles (Peters and Hewlett 2006, 93). Ironically, at least on the academic front, fundamentalists did not initially oppose evolution (Peters and Hewlett 2003, 122). Over time, that attitude changed, and I suspect that for fundamentalists at the popular level, a literal understanding of the six days of creation was part and parcel of the belief in the inerrancy of scripture.

The Scopes Trial in Tennessee in 1925, as we have mentioned before, is taken in the popular mindset as the classical example of the religion-science conflict. It is certainly depicted as the classical image of the confrontation between fundamentalism and secular science. It is also the culmination of the battle between the modernist (liberals)-fundamentalist battles in the mainline denominations, won by the liberals. What is virtually forgotten is that John Scopes, the young teacher who defied the laws of the state of Tennessee by teaching evolution, did so, encouraged in this endeavor by his friends at church, as a devout Methodist who embraced modern science.

Since I have been talking about fundamentalism, I would like to bring some greater precision to my use of terms. Although, to be sure there is a long if not dominant history of literal interpretation of the Bible in Christianity, fundamentalism, as we have seen, is a very modern phenomenon. At one level, it is part of modernity's quest for absolute certainty. At another level, it is, as Langdon Gilkey maintained in numerous writings, a *partial* rejection of the modern world—and especially of

Liberal Protestantism's embrace of that world. For example, following the Scopes Trial, although Scopes was found guilty and fined, fundamentalists, regardless of the number of their adherents, "circled the wagons" and withdrew from the public square and popular culture until the 1970s. Since then, we have seen a resurgence of fundamentalism, seeking absolute certainty with a literalist biblical interpretation, rejecting at least aspects of modern science and most especially its attendant secularity yet making the most adept use of the latest technology, public relations, and political communication.

No less a figure in conservative Christianity than Billy Graham drew a distinction between fundamentalism and evangelicalism: "All fundamentalists are evangelicals but not all evangelicals are fundamentalists." Fundamentalists are literal about the Bible whereas evangelicals, while conservative in their interpretation, are not necessarily literalists and by and largely accept the findings of modern science, with many accepting the theory of evolution. Fundamentalists and non-literalist evangelicals alike stress the importance of a deep, transformative experience, the "born again experience."

An interesting similarity between evangelicalism and at least nineteenth, early twentieth-century Liberal Protestantism is a concern with what is the "essence" of Christianity vs. the "non-essentials" (today's Liberal Protestants are prone to think that there is no essence to anything). In Liberal Protestantism, this essence was minimal, love of God and one's fellow humans, seeking the Kingdom of God with some measure of social justice. For evangelicals, the "essence" of Christianity is much broader, perhaps including the "born again" experience, a conservative though not literal reading of scripture, and some standards of conduct. Both are concerned with the question of Christian identity with evangelicals willing to concede far less to the modern world.

One last point in this discussion concerns the use of the word "modernist." The word is typically used in relation to the Roman Catholic Modernists of the late nineteenth, early twentieth centuries whose works were condemned for their "modernism." In Roman Catholicism, modernism did not apply so much to the embrace of modern science as to the use of the tools of historical research toward the Bible and church documents and a developmental view of church history. The Liberal Protestants of the period we are covering used terms liberal and modernist interchangeably. Some of today's commentators treat that era's Protestant modernism as a forerunner of those who today deny that there is an essence to any religious tradition.

In spite of some rhetoric that resembles that of conservative Christians, I would now like to turn to what amounts to a family squabble in the camp of Liberal Protestantism.

THE NEO-ORTHODOX INTERLUDE: THE INDEPENDENCE OF SCIENCE AND RELIGION

On June 28, 1914, the Archduke Franz Ferdinand, heir to the throne of the Austro-Hungarian Empire, was assassinated by the Serbian national-ist Gavrilo Princip. The unimaginable happened: all of the great powers declared war on one another in a great worldwide conflagration. The war became a protracted conflict fought largely in trenches that barely moved as one army advanced, the other retreated. Given the ethos of optimism and confidence in progress of the time, this was not supposed to happen! The war ended with the collapse of four great historic empires, urban strife, and revolution.

Within the Liberal Protestant camp, at least in Europe, there was dis-illusionment. The confidence and optimism of Liberal Protestantism was inadequate to deal with war, social injustice, and revolution. A symbol of this inadequacy was the figure of Adolf von Harnack (1851–1930), per-haps Germany's leading Liberal Protestant theologian, writing the German Emperor, Kaiser Wilhelm's declaration of war!

The leading voice in articulating this disillusionment was that of Karl Barth (1886–1968). Coming out of the liberal camp, having been one of the students of Wilhelm Hermann, Barth served a church in a Swiss mining town during the war. He came to feel that the liberal gospel of which he was an enthusiastic devotee simply did not speak to the exploited lives of the miners nor did it speak to the situation of war. It was out of this felt need that Barth published *The Epistle to the Romans* in 1919.

The commentary shook the ground under the feet of Liberal Protes-tantism. Barth went back to the Bible and to the classical affirmations of the Reformation in order to let the Gospel speak. Thus, instead of stress-ing the imminent God of evolutionary theism, he affirmed a God who is Wholly Other. Denying that continuity between God and God's creatures, he maintained that God could not be known through reason and experi-ence; she/he could be known *only* because of God's decision to reveal the divine self in Jesus Christ.

This kind of outlook did not leave much room for a dialogue between religion and science. In fact, Barth drew a sharp distinction between reli-gion and the Christian faith. Religions are made by humans and as such are idolatrous while the Christian faith is revealed by God. While science has its proper realm, it does not have anything to say to the Christian faith and theology. Science is a totally human construct as well.

Barth did believe in the findings of modern science as well as in the practical effects of science. But to him, the realms in which science and theology operated were entirely different and ought to be kept separate. Christian theology, "faith seeking understanding," had its own norms.

Figure 4.3 Karl Barth (Courtesy Library of Congress)

In a fashion similar to his attitude toward science, Barth accepted the use of the tools of historical-literary criticism when it comes to the Bible. Nevertheless, on a trip to America, when he was asked what was the "core" of his Christian faith, he replied, "Jesus loves me this I know, for the Bible tells me so!" American fundamentalists claimed Barth rather inaccurately as one of their own!

His theology came to be called "neo-orthodox," "new orthodox." It was still "orthodox" *but* it was new in accepting the premises of liberal theology. However, instead of the radical and enthusiastic embrace of modern science and reinterpreting the Christian faith in its light, it saw faith and science needing to live in totally separate realms.

Although he eschewed the designation "neo-orthodox" and was quite critical of Barth, Reinhold Niebuhr (1892–1971) was the greatest representative of neo-orthodoxy and its disillusionment with liberalism in the United States, although, like Barth, within the liberal camp. Since it was not fought on American soil, World War I did not quite have the dramatic impact on liberal optimism in the United States that it had on the continent. Large-scale disillusionment did not set in the culture at large until the onset of the Great Depression in 1929. For Niebuhr, the disillusionment set in during his fourteen years as pastor of a church in downtown Detroit, where he experienced industrial strife first hand. Like Barth before him, Niebuhr felt that Liberal Protestantism was inadequate to the challenges of the twentieth century.

In 1929, he accepted a position as Professor of Social Ethics at Union Theological Seminary in New York City. He continued an intense search for what he thought was the "core" or most important aspect of Christianity that spoke to the people and issues of the day. Never dogmatic, he had become a socialist in Detroit and ran as a Socialist for Congress in 1930 in New York City. He supported the Socialist Party's nominee, Norman Thomas, for president in 1932 and 1936.

In his writings, Niebuhr continued this combination of a search for what was most distinctive about Christianity that spoke to and illuminated the issues of the day. Again, never dogmatic, in his early works he made use of Marxian analysis. In his *Moral Man and Immoral Society* (1932), he was already stressing the radical nature of human sin, self-centeredness. This radical self-centeredness is aggrandized in nations and communities. As a result, different ethical norms apply to individuals and interpersonal relations than between nations and communities: love is an ethical norm applied to individuals and in interpersonal relations while in the relation between nations and communities, given their aggrandized collective self-regard, the only possibility is an ever changing, ever shifting scale of justice. In his *Introduction to Christian Ethics* (1935), human are so radically self-centered that love in the sense of "agape," self-giving, self-sacrificing love is an impossibility in human affairs. The only kind of love possible in human affairs is mutual, reciprocal love, characterized by give and take but always tainted by sin, self-centeredness. The only possibility for "agape" to be effective in the world is as a transcendent principle that judges all partial accomplishment of justice.

It is in his Gifford Lecture, published as *The Nature and Destiny of Man,* that Niebuhr sets forth his mature view of what is distinctive about the Christian faith. In order to find out what that is, he engages in a lengthy analysis of human existence. Influenced by existentialism, with which we shall deal later, Niebuhr sees humans caught at the juncture between finitude and freedom or, alternatively, nature and spirit. Finitude, synonymous with nature, has to do with our limitations, linked to our "bodiliness." Freedom, synonymous with spirit, has to do with our capacity to transcend ourselves.

We are caught in the middle, at the juncture of finitude and freedom, nature and spirit, and we are not very good at handling it. Niebuhr uses the simile of the sailor climbing toward the mast above and staring at the abyss below. This for him is the human condition.

Living at the juncture of nature and spirit, finitude and freedom, climbing toward the mast above and staring at the abyss below, makes us anxious. All forms of anxiety, from seeking a better grade to being on time to trying to impress the person one likes, are symptomatic of anxiety—the threat of the loss of the meaning of existence. To deal with this polarity, human beings tend to deny one side of the nature/finitude—spirit/freedom polarity in favor of the other. Thus, some people get so caught up in the vitalities of life that they deny their capacity for self-transcendence and for taking historical responsibility. This is what Niebuhr calls the sin of sensuality. Although there are people who fall into the sin of sensuality, he thinks there are more who fall into the sin of pride, which is the denial of our finitude and consequent seeing ourselves strictly in terms of self-transcendence and freedom. The result is an undue self-regard by which we put ourselves at the center of the universe. We can see the devastating effects of the sin of pride if we analyze the phenomenon of love again.

For Niebuhr, agape is a quality of God's love alone; only God is capable of the totally self-sacrificing quality implied by the word agape. The best that we can hope for is "philia," brotherly/sisterly love, a mutual, reciprocal love of give and take. This is an important kind of love, always qualified and relativized by agape, which stands as a transcendent norm of judgment. However, mutual love, in Niebuhr's estimation, is profoundly tainted by sin; no matter how much our altruistic, self-giving love may be, there is always something in it for us.

It is even worse in the case of nations and communities. As he also states in *Moral Man and Immoral Society*, laws of love do not apply to nations and communities. The undue self-regard of human beings is magnified and aggrandized in the lives of nations and communities so as to make justice the form love takes in their relations with one another. Thus, the best that can be hoped for is an ever shifting, dynamic, balance of power.

In spite of this, just as agape is the transcendent norm of judgment that stands over mutual love—and justice for that matter—so stand the partial approximations of mutual love over justice. Thus, there is a dialectical relationship among philia, and justice. In similar fashion, Niebuhr maintains that, the perennial principles of justice, freedom, and equality function as transcendent principles judging all partial approximations of freedom and equality. The principles of freedom and equality ever serve to prod us into ever changing, dynamic further realizations of those perennial principles.

For Niebuhr, there is a need for balance in every aspect of human affairs. Thus, freedom and equality need to be in delicate balance. In similar fashion, the twin threats to communal existence are tyranny and anarchy. Communities do need some degree of centralized power to exist and to establish some modicum of justice, otherwise chaos takes over. On the other hand, there needs to be some system of checks and balances in order to prevent undue concentrations of power in the hands of both individuals and institutions.

Reassessing his previous doubts about President Franklin Delano Roosevelt, by 1940 Niebuhr came to think that the New Deal represented a pragmatic experimentation that sought a balance between extremes, as between freedom and equality, and respected checks and balances. He supported Roosevelt in 1940 and 1944. Later, in his *The Irony of American History* (Niebuhr 1952), he would uplift the ever shifting American system of checks and balances as prototypical of the kind of system requisite and necessary to prevent undue concentrations of power.

The widespread public influence of Reinhold Niebuhr cannot be minimized. In 1946, he served on the State Department's Policy Planning Council. He participated in the formulation and implementation of George F. Kennan's "containment" policy, which maintained that the purpose of U.S. foreign policy was to "contain" the Soviet Union. He was one of the founders, with such friends as Hubert H. Humphrey, of the staunchly New Deal liberal and equally staunchly anti-Communist organization, Americans for Democratic Action. Perhaps the greatest indicator of the pervasiveness of his influence was the appearance of his picture on the cover of *Time* in 1948 and his being touted in this manner as a potential presidential candidate by the magazine's staunchly anti-Communist Republican founder Henry R. Luce.

Niebuhr's influence in the academy, foreign policy circles, and international theory was perhaps greatest in propagating the notion that public policy was predicated on a certain view of human nature. His view of human nature was fundamentally shaped by the non-literal understanding of original sin, which to him was Christianity's distinctive contribution to the issues confronting the modern world. He called his understanding of human nature and of the world "Christian Realism." Others in the "realist" school of international relations may have foregone the designation "Christian," but they agreed with him that there needs to be an adequate understanding of human beings underpinning international relations. And that view was that humans were fundamentally self-centered, pursuing their self-interest. Agreeing with Niebuhr, they maintained that humans have a remarkable capacity for self-deceit about their self-interest, attributing altruism to one's actions or those of one's group when in fact those actions were tainted by egoism. Some realists agreed with the Christian realist theologian that there were perennial ethical principles, such as freedom and equality, at play in the lives of nations and communities. Others, such as the extremely influential Hans Morgenthau, did not.

Niebuhr was the quintessential Cold Warrior after 1946. However, he was never under the illusion of the innocence of the exercise of American power and consistently pointed to the self-deceit manifest in aspects of American foreign policy. There were vital interests and fundamental values at stake in the Cold War. But that was never an excuse for not exercising the capacity for self-examination and self-criticism; power has to be exercised responsibly. The quintessential Cold Warrior became a staunch critic of the Vietnam War in the last decade of his life.

My readers may rightly wonder what Niebuhr and his theological ethics has to do with the relationship between religion and science. My answer is twofold.

First, Niebuhr uses his "realistic" estimate of human nature to critique the liberal optimism about the possibilities of human existence. He uses that realism to critique the "myth" of evolution, not in the popular sense of a lie but in the sense of a profound orientation toward life and history behind the literal reading that searches for absolute certainty as it contributes to what he thinks is Liberal Protestantism's unwarranted optimism about human nature.

The second concern is fundamentally related to the first. Niebuhr is concerned with the ethical uses of modern science. He was especially concerned with the practical effects of scientific discoveries, such as the development of nuclear weapons.

On the whole, however, I would contend that the effect of Niebuhr's critique of Liberal Protestantism's radical embrace of modern science (which he embraced no less, at least in terms of accepting the factuality of its findings), was to contribute to the "neo- orthodox interlude." That is, he helped shape the sensibility that science and religion deal with two different realms of truth that exist side by side.

Lest I paint a one-sided picture of Reinhold Niebuhr as a representative of neo- orthodoxy, I need to point out that he eschewed that designation. Furthermore, such process theologians as Bernard M. Loomer, Daniel Day Williams, and Bernard E. Meland (who while using the insights of process thought did not want to be considered a process theologian), whom I shall consider later, were all in various ways influenced by him. Loomer appropriated the category of "ambiguity" to describe existence while Williams developed an appreciative critique from a liberal perspective that nevertheless did not deny Niebuhr's central insights (Williams 1949).

Before we take a look at the relationship of religion and science in the creative theology of Karl Heim, who was a Christian existentialist, we need to look at the meaning of existentialism. Existentialists are notorious for resisting precise definitions of "existentialism" and "existentialist." They feel such precision is too "objectifying," depersonalizing in a world in which human individuals are already treated like objects, numbers instead of persons, in which concrete individuality is swallowed up by philosophical and socio-politic-economic-cultural systems.

Nevertheless, we need to speak to one another. Thus, while resisting being pigeon holed by precise definitions, existentialism and its view of human beings can be described as having certain characteristics. One example is that all existentialists emphasize and employ the term "existential" to refer to that which is most "here and now," that which is most intensely personal, that which is of the most immediate concern to me. At one level, this can be anything: I may be concerned about my relationship to my "significant other," something I said to my child and the manner in which I said it, the test tomorrow or something as seemingly mundane as going to the bathroom. But human existence is also characterized by certain pervasive themes and questions.

One persistent theme in existentialism is death. To borrow from Martin Heidegger (1889–1976), life is a train ride toward death. Humans beings, as far as we know, are the only creatures aware of our mortality. We are not very good at handling this knowledge and the ensuing anxiety. And yet, there is much more! Human beings are also radically free to choose who they are in every moment. Combine this awareness of our radical freedom with consciousness of our mortality—the result is Anxiety (the similarities to Reinhold Niebuhr should be rather apparent). We do not like that state of Anxiety and we do not know how to handle our freedom and the awareness of our mortality. Thus, we are confronted every moment (every moment being discrete, isolated, completely novel, disconnected from every other moment) with having to decide to live authentically, to be ourselves, or inauthentically, allowing some external source to dictate who to be.

There are religious as well as atheistic forms of existentialism. Atheistic forms of existentialism, such as that of Jean-Paul Sartre (1905–1980), maintain that authentic existence is choosing to live as oneself every moment. Recourse to anything outside of oneself, whether it be God, peer pressure, totalitarian movements, the tastes of popular culture, is to fall into the trap of inauthenticity. In some forms of existentialism, such as the philosophy of Karl Jaspers, as one decides to live authentically, one is in touch with the Transcendence that is basic to reality. In the thought of the Jewish existentialist Martin Buber (1878–1965), authentic existence is engaging in "I-Thou" relationships, with humans and non-humans alike, subject to subject, instead of perpetuating the endless cycle of impersonal "I-It," object to object relationships that seem to characterize our age. In the theology and New Testament interpretation of Rudolf Bultmann (1884–1976), which appropriates the existentialism of Martin Heidegger, the decision we face every moment whether to live authentically or inauthentically is a decision to accept, reject, be indifferent toward the grace of God in Jesus Christ, the unmerited, unbounded, unconditional love of God in

Jesus Christ that confronts us, accepts us as we are, and empowers us to live authentically.

The epistemology of existentialism, an heir to the traditions of Descartes and Kant, is profoundly dualistic. There two are two kinds of knowledge. First, there is "objective" knowledge, the kind of supposedly dispassionate knowledge gained through the scientific method. But there is a different, more profound kind of knowledge as well. This kind of knowledge is "subjective," personal, passionate—it is, in other words, "existential." Even if the latter is privileged, both kinds of knowledge are valid in their own separate realms.

In spite of coming from the side of the neo-orthodox tradition that made use of existentialism, Karl Heim (1874–1958) tried to overcome this impasse. He was one of the first theologians to note that, unlike the thinkers of the Enlightenment, for whom there was increasingly little for God to do, God's existence was nevertheless presupposed. The radical secularism of at least the Western world, especially Western Europe of the mid-twentieth century (in his historical context, it was much more typical to generalize and to universalize!) had left no room for God. This secularism was tied to the mechanistic view of the world described in Chapter 2. In this secularist vision, the idea of God, is at best irrelevant. The world of the secularist and the world of the believer indeed seem to be in different realms, light years from each other.

Heim, however, found certain developments in modern science, especially modern physics, to provide an opening for a different way of relating science and religion. Modern physics discovered that everything is energy, that everything in the world is a different form of energy and its self-organization. The idea that everything is energy along with the idea of the complementarity of the wave-particle character of light (rather the a clash of opposites), shows that reality is one; knowledge of it cannot be neatly compartmentalized. Finally, the emphasis on energy opened up new ways of talking about a trans-spatial reality, the omnipresence of God.

Heim does not walk completely through the door he had opened. He stills ends with a world in which science and religion deal with different realms: " ... we are faced ... with a conflict between two possibilities which cannot be reconciled" (Heim 1957, 247). He writes:

The first possibility is the general outlook of secularism, that with which we all find ourselves at birth ....The second possibility is the world-picture of faith, a picture in which the personal God forms the supreme center ... We are not born into this second overall view of the world; it can only fall into our laps by means of a "second birth," as a gift which we cannot ourselves procure but which, when it has been apportioned to us, we also cannot ourselves revoke. The conflict between these two general conceptions, from which there also always result two contrary views of nature, is the deepest tensions in which we live, including the ideological struggles which affect even political and economic life. (Heim 1957, 247)

He maintains further:

For we stand in two spaces at once, spaces with contrasting structures. The one is the space into which we have been born, together with all other beings. In this space we live and think and explore in accordance with the generally accepted methods of natural science together with all the others. We can communicate with all thinking beings in a way which ensures mutual comprehension and general agreement. The second space is that which is disclosed to us only by a "second birth," as it were by a "second sight." With regard to this second space we can communicate only with those who have undergone the same experience as ourselves. Even we ourselves can live in this second space only by ever anew wielding the "nevertheless of faith" to combat whatever comes us from the polar space and tires to cast doubt upon the reality of suprapolar polar space. This cleavage, which runs right through the whole state of the world, does not merely prevent the ending of the ideological struggle between the secularists and those rooted I religion; it passes though the soul of the believer himself as a daily temptation to forsake his faith. (Heim 1957, 248–249)

Thus Heim, to use the language of Barbour and Haught, still winds up with neo-orthodoxy's independence and contrast models of relating religion and science.

The neo-orthodox interlude seems to be quite distant from the enthusiastic embrace of modern science that we have seen in the earlier chapters of this volume. Yet, it is extremely important to remember that all of the neo-orthodox theologians took the truths discovered by modern science for granted. Their fight with Liberal Protestantism had the character of a family fight.

Chapter 5

<hr>

Science, Liberal Protestantism, and the Twentieth/Twenty-First Centuries: 1960–2006

THE DIALOGUE RESUMES

Even during the heyday of neo-orthodoxy, there were places and individuals that kept alive the tradition of the Liberal Protestant enthusiastic embrace of modern science and perhaps channeled it in a different direction. One such voice was that of Henry Nelson Wieman (1884–1975).

After finishing his Ph.D. at Harvard, Wieman, who had been ordained a Presbyterian minister in 1912, taught at Occidental College from 1917 to 1927. In 1926, he was invited to give a lecture at the University of Chicago Divinity School on Alfred North Whitehead's *Religion in the Making*. The lecture was such a resounding success that he was offered a position beginning the following year, an offer that he accepted. He stayed at Chicago until his retirement in 1947. Following his retirement, he taught in numerous places, perhaps most notably at Southern Illinois University.

Standing within Chicago's liberal tradition, Wieman would help to take it in new directions. The older "Chicago theologians," while certainly influenced by the natural sciences and evolutionary theory, were more interested in using the social sciences to study Christianity as a dynamic, creative socio-historic movement. Wieman, on the other hand, was instrumental in inaugurating a different, far more philosophical course that appropriated the insights of the natural sciences, although, to be sure, he did make extensive use of psychology.

Throughout his works, Wieman is haunted by the question of what operates in human existence to create, sustain, save, and transform toward the greatest good and what are the conditions which must be present for its most effective operation? His interest, as we shall discuss at greater length below, is to develop an empirical theology.

He also seeks to develop a theology that is strictly naturalistic: there is no world beyond this one, and God is in this world or nowhere. He claims that there is nothing that we can know except events, their qualities, their relations, and their possibilities (Wieman 1967, 6–10). If that is so, there can be no reference to anything transcendental or supernatural.

Wieman develops his central theological tenet, "the creative event," which he later came to call "creative interchange," in the context of a discussion of value. Wieman describes his understanding of value in the following manner:

We shall try to demonstrate that there is a creative process working in our midst which transforms the human and the world relative to the human mind. We shall then show how transformation by this process is always in the direction of greater good. The human good thus created includes goods, satisfactions of human wants, richness of quality, and power of man to control the course of events. But the greater good cannot be attained by seeking directly to increase goods or satisfactions or quality or power. These can be increased only by promoting that kind of transformation creative of the greater content of good when created good is interpreted as qualitative meaning. (Wieman 1967, 17)

Qualitative meaning is *created good* whereas "... there is a prior kind of good here called *creative*, which alone is the source of life's abundance" (Wieman 1967, 17). He further defines qualitative meaning as "... that connection between events whereby present happenings enable me to feel not only the quality intrinsic to the events now occurring but also the qualities of many other events that are related to them" (Wieman 1967, 18). Moreover, "qualitative meaning is that connection between events whereby the present happening conveys to me to feel qualities of other happenings and some qualities pertaining to what will happen in the future, as the future is interpreted by the past (Wieman 1967, 18–19).

Although qualitative meaning is intrinsically good and the increase of good is correlative with the increase of qualitative meaning, it is not the "guiding thread" that Wieman seeks. This "guiding thread" he finds in the source of qualitative meaning, which he calls the "creative event." He maintains that "the creative event ... weaves a web of meaning between individuals and groups and between the organism and its environment" (Wieman 1967, 20). While humans cannot control the creative event, there is much they can do to enable or obstruct the conditions for the release of the creative power of the event (Wieman 1967, 20). Wieman asserts that one of the most important of these conditions is that the individual give herself/himself to this process of creative transformation, for " in weaving the web of richer meaning, the creative event transforms the individual person so that he is more of a person" (Wieman 1967, 20).

Wieman describes the creative event in terms of four subevents. The first of these subevents is "emerging awareness of qualitative meaning derived from other persons through communication," certainly biologically based and fundamental to our becoming human (Wieman 1967, 59). The second is the integration of "... these new meanings with others previously acquired," with the thoughts and feelings deepened (Wieman 1967, 58–59). This integration usually takes place in solitude although its insights are eventually communicated, whether through speech or the written word (Wieman 1967, 60). The third subevent is "expanding the richness of quality in the appreciable world by enlarging its meaning" (Wieman 1967, 58). There is a new structure of interrelatedness that expands and enriches the appreciable world as a result of the first two subevents. There is a new richness and variety in one's experience (Wieman 1967, 61–62). This may lead to greater loneliness, making one aware of a greater good that may never be achieved, a greater love that may never be requited (Wieman 1967, 62–63). The third subevent is "... an expansion of the individual's capacity to appreciate and his apprehension of a good that might be, but is not fulfilled" (Wieman 1967, 63).

The fourth subevent is the "widening and deepening community between those who participate in the total creative event ..." (Wieman 1967, 64). Wieman writes:

The new structure of interrelatedness pertaining to events, resulting from communication and integration of meanings, transforms not only the mind of the individual and his appreciable world but also his relations with those who have participated with him in this occurrence. Since the meanings communicated to him from them have now become integrated into his own mentality, he feels something of what they feel, sees something of what they see, thinks some of their own thought. (Wieman 1967, 64)

He describes this widening of community further:

This community includes both intellectual understanding of one another and the feeling of another's feelings, the ability to correct and criticize one another understandingly and constructively. It includes the ability and the will to co-operate in such a manner as to conserve the good of life achieved to date and to provide conditions for its increase. (Wieman 1967, 64)

The life, death, and resurrection of Jesus provides Wieman's example of the creative event and the working of the four subevents. Through interaction with Jesus and the consequently different kind of interaction with each other, they became different people, the thoughts and feelings of each making a difference to each and to all (Wieman 1967, 39–40). The important thing was "not something handed down to them from Jesus but something rising up out of their midst in creative power . . ." (Wieman 1967, 40). This was not anything particular Jesus did but rather ". . . something that happened when he was present like a catalytic agent" (Wieman, 1967, 40). Wieman states that "it was as if he was a neutron that started a chain reaction of creative transformation" (Wieman 1967, 40).

Moreover, illustrating the second subevent, new meanings were integrated with the old. Each became more sensitive and appreciative, something that led right into an incarnation of the third subevent. The "appreciable world" of each disciple expanded. Because of this, the appreciable world of the community expanded as well. This in turn led to a fourth consequence, the fourth subevent—". . .more depth and breadth of community between them as individuals and between them and all other men" (Wieman 1967, 40–41). Wieman states eloquently:

What happened in the group about Jesus was the lifting of this creative event to dominate their lives.

What happened after the death of Jesus was the release of this creative power from constraints and limitations previously confining it; also the formation of a fellowship with an organization, ritual, symbols, and documents by which this dominance of the creative event over human concern might be perpetuated through history. Of course, there was little intellectual understanding of it. . .(Wieman 1967, 41)

He describes the resurrection in the following manner:

After about the third day, however, when the numbness of the shock had worn away, something happened. The life-transforming creativity previously known only in the fellowship with Jesus began again to work in the fellowship of the disciples. It was risen from the dead. Since they had never experienced it except in association with Jesus, it seemed to them that the man Jesus himself was actually present, walking and talking with them. Some thought they saw him and touched him in physical presence. But what rose from the dead was not the man Jesus; it was creative power. It was the living God that works in time. It was the Second Person of the Trinity. It was Christ the God, not Jesus the man. (Wieman 1967, 44)

As I have maintained before, Wieman seeks to be a naturalist and empirical as a theologian. Clearly, he wants to establish the working of the creative good on strictly naturalistic grounds, i.e., with no recourse to anything above or beyond the natural world. At this point, we need to take a more in depth look at the meaning of the word "empirical."

First of all, the words "empirical," "empiricism" refer to "experience," knowledge gained though experience. British empiricism and reductionistic scientific materialism restrict the meaning of experience to sense experience. Radical empiricism, on the other hand, claims that, while sense experience is usually a dimension of most experience, there is a depth, range, and intensity to experience that cannot be restricted to sense experience or quantified. Thus, experience is primarily the experience of feelings. While sense experience usually accompanies most experience, it is secondary and derivative. Feelings are more than the manner in which the chemicals interact in our bodies, although they are, to be sure, that as well.

Radical empiricism resembles Romanticism in considering feelings to be the primary means of knowing, whether in the case of human beings or anything actual at all (Barzun 1976, 1, 9, 133; Randall 1976, 400). If we consider the ideas of William James and Alfred North Whitehead two of the most prominent advocates of radical empiricism (James being the one who coined the term), we find that for both the basic unit of reality is a momentary experience, a "puff of experience" as James calls it, an "actual occasion of experience" for Whitehead. Momentary experiences, human and non-human, constitute themselves as they "feel" the "feelings" of the objective world and reach out for the possibilities of the future. For both James and Whitehead, feelings are primordial, comprising what James calls "pure experience."

In the case of both of these thinkers, reason is a particular feature of experience, human and nonhuman.

By seeing reason as a highly complex dimension of experience, that is to say feelings, radical empiricism's non-sensationist understanding of perception seeks to overcome the traditional dualism between emotion and reason. In similar fashion, radical empiricism can be viewed as attempting to overcome the dichotomy between rationalism and empiricism.

Radical empiricism shares with Romanticism an organismic understanding of the non-human natural world. This is certainly true of Whitehead. James, on the other hand, has been interpreted by some to be individualistic and concerned solely with human beings. While human individuals are admittedly James' main concern, I would maintain that those concerns need to be seen within the broader contours of his thought. By this I mean to say that for this representative of Classical American Philosophy, the individual momentary experiencing, "puffs of experience," human *and* non-human, arise out of and contribute to a web of relationships, human *and* non-human, in a pluralistic universe constituted of interdependent parts (Ford 1982, 75–107; Ford 1993, 99–105). Because experience, human *and* nonhuman, in all its depth, range, and intensity is the locus of value, radical empiricism sees all experience, human *and* non-human, as having some degree of intrinsic value.

Like Romanticism, radical empiricism extols both the individual and the community, "the individual-in-community." We have already seen that for radical empiricism the individual emerges out of a matrix of relationships, human and non- human, of which she/he is a part and which is a part of her/him. Moreover, another dimension of this emphasis on community can be seen in radical empiricism's claim that what we "feel" in the act of knowing are relations as we bring the multiplicity of data in our past, as well as the past of the entire universe, into some sense of unity or creative synthesis in the subjective experiencing of the moment.

The previous discussion raises the question of whether radical empiricism is truly scientific. It goes without saying that to reductionistic scientists, who restrict the meaning of knowledge to what we can know through sense experience, radical empiricism is not scientific. Instead of the value-free objectivity so highly prized by modern science, radical empiricism introduces elements of subjectivity that are notoriously unreliable and outside of what they see as the proper bounds of scientific inquiry.

Another way of dealing with the issue is to separate questions of scientific methodology from questions of philosophy, epistemology, and metaphysics. This view maintains that science is strictly a method from which no particular conclusions about the way(s) we know or views about the nature of reality can be drawn. Thus, there are strict boundaries drawn between scientific and philosophical questions.

A third view embraces the notion that radical empiricism is not unscientific. Indeed, this was the view of James and Whitehead as well as many of their followers. In this regard, it might be important to point out that both James and Whitehead started with scientific backgrounds, James in medicine, Whitehead in mathematics with a profound interest in physics, including the theory of relativity.

According to this view, if the primary way we know is through feeling in the subjective experiencing of the moment, science needs to take this into account. For one thing, the engagement of many scientists in scientific inquiry is motivated by their subjective sense of wonder at the universe. Moreover, especially with the influence of quantum theory and ecology, there is an increasing recognition that there is a subjective element involved in scientific knowing. Many scientists acknowledge that there is a subjective element involved in the subject matter chosen for an experiment as there is in determining the results and drawing the conclusions, especially in such notions as that of the "participant observer." Deconstructive postmodernism has prompted discussions of science as a social construction.

In the previous discussion of radical empiricism, I have presented some basic tenets of Whiteheadian process thought that had an influence on Wieman's early work. At this point, I would like to finish fleshing out some other concepts in process thought critical to the theme of this book. I have to confess that I have a tendency to read James through my Whiteheadian lens, Whitehead through my Jamesian lens.

In describing the subjective experience of the moment, Whitehead maintains that all actualities, from the tiniest energy to human beings, strive for the experience of beauty. There are two components in the experience of beauty: harmony and intensity. In order, for there to be some semblance of beauty, there needs to be harmony. However, there can be too much harmony, in which case experience becomes trivialized. Life can become boring—just think of the numerous complaints by people about their boring jobs! Hence, there is the need for contrast, for intensity of feeling. Of course, it is possible to have too much intensity. Life becomes chaotic and overwhelming. Thus, beauty is the dynamic and delicate balance between harmony and intensity.

In Whiteheadian process thought, all actualities, human and non-human, strive to realize themselves in fundamental interdependence with one another. It is at this point that Whitehead introduces the idea of God.

God, in the "primordial nature" or what I like to call the active side of God, first of all God envisages *all* possibilities. The possibilities we encounter are not infinite and chaotic but have some sense of order or graded relevance to who we have become and the decisions we have made in the past. For example, at 5 feet $3^1/_2$ inches I am not likely to grow to 7 feet in order that I play Shaquille O'Neill one on one! What is likely is that I can enjoy the weekend in a way that is congruent with the person I have become, become a better person, a better teacher.

If there is an order to the possibilities we encounter, it stands to reason that there is an "orderer." That orderer is God, who in the primordial nature or active side, orders possibilities in graded relevance to the experiences and past decisions of all actualities. More than that, God "lures" all actualities to realize themselves not egotistically or in isolation but in fundamental interdependence with each other. Like a number of the Liberal Protestant theologians we have considered earlier, Whitehead rebelled against tyrannical notions of God. Hence for him, God always acts "persuasively" not coercively, not in the sense of rational arguments, but luring with ideals to be actualized.

God is not an exception to metaphysical categories but their chief exemplification. If that is so, then just as all actualities receive data from the past, not just their own but from the entire universe, God receives data from all actualities. God, in her/his "consequent nature" or receptive side experiences all experiences in the world and preserves them everlastingly with no loss of immediacy. Thus, while each subjective moment of experiencing has significance, dignity, and intrinsic value in its own right, there is ultimate significance and dignity bestowed on each moment by being preserved in the divine life, creatively transforming "the perpetual perishing" that occurs each moment.

Wieman, although influenced by Whitehead early in his career (he had also been influenced by the French philosopher Henri Bergson [1859–1941]), thought the idea of the consequent nature of God was totally speculative and unempirical and eventually distanced himself from Whitehead and his followers although he retained a process metaphysics. He was increasingly influenced by John Dewey (1859–1952) as he got older. Dewey, no less than James and Whitehead, defined radical empiricism in terms of the breadth, depth, and intensity of experience. Dewey, nevertheless, was an instrumentalist, concerned with how a thing functioned, how one could observe it functioning, and the very practical results of how it led to the resolution of concrete problems. It was this aspect of Dewey's empiricism that Wieman appropriated. Although Wieman is keenly aware of the element of interpretation, as in his description of the life, death, and resurrection of Jesus as the model for the creative event, his analysis of what constitutes human good, the creative event, and the fourfold subevent are "empirical" in Deweyan sense just described.

Wieman's former student and onetime collaborator Bernard E. Meland (1899– 1993) in the 1930s, who taught at Pomona College for a time and then came back to teach at the University of Chicago School Divinity School in the 1940s, continued the radical empirical tradition, appropriating James and Whitehead. Also a student of Gerald Birney Smith, Meland initially sought to develop his mentor's "mystical naturalism," a sense of humanity's earth creatureliness, based on modern science. Over time, other concerns came to the forefront. Standing firmly within the tradition of Liberal Protestantism, using the radical epistemologies of James and Whitehead, Meland, like Wieman, also attempted to answer the concerns of neo-orthodoxy. Meland often used expressions such as "we live more deeply than we can think" and "a good not our own" in speaking of God, clearly emphasizing the difference between God and the creatures. Wieman does something similar in his distinction between the created good and the creative good, which is a good not our own that we cannot control, only be open to or obstruct its operation. Meland, in spite of the influence of Whiteheadian process thought, nevertheless avoided speaking in too precise terms about God, preferring to speak of " Mystery" or "Creative Passage." It needs to be pointed out that he used the epistemology of radical empiricism to engage liberal theology with modern science, tempering Protestant Liberalism's easy going optimism in the process.

There were other significant figures during this period in the Liberal Protestant environment of Chicago. One was Bernard M. Loomer (1910–1985), who became Dean of the University of Chicago Divinity School in 1943 at the ripe old age of thirty-three. He tried to guide the Divinity School into the unifying vision of Alfred North Whitehead. Although as dean he indulged in the speculative side of Whiteheadian process thought, Loomer stands squarely in the empirical side of the process tradition.

In the 1970s and 1980s, he worked on the notion of the "size" of God. That is to say, if the experience of beauty entails intensity, the ability to take more and more of the world into oneself, at least by analogy, the same notion is applicable to God, that is, God grows in the divine stature as God experiences all experiences. So far, Loomer is following the Whiteheadian argument. However, he parts company with Whitehead and most process theologians in arguing that if God experiences all experiences, then God experiences the ambiguity of the world. And if God experiences the ambiguity of the world, then God, instead of being all good, is ambiguous. Loomer came to a pantheist position, identifying God with the web of life.

All of this may seem terribly speculative compared to the supposed empiricism of science. However, Loomer's work discloses the inherent interest process thought has in science and the influence of the discipline of ecology. Moreover, if we use the definition of empiricism in the sense of radical empiricism, of experience not being limited to sense experience but including the depth, intensity, and richness of experience, all in a creative web of life in which the diverse parts are interrelated, then we can see the influence of modern science, particularly ecology. "Life" itself is characterized by capacity to be sensitive to one's world, to respond, to take more and more of the world into oneself.

Daniel Day Williams (1910–1973) was another liberal theologian of note whose work spanned several theological eras. A United Church of Christ minister (at that time called Congregational) who had served churches in Colorado, Williams accepted in the early 1940s a professorship at Chicago Theological Seminary, which was then part of Federated Faculty with University of Chicago Divinity School and other Chicago area seminaries. He moved on to Union Theological Seminary in 1955. He was initially influenced by the empiricism of Wieman, as is evident in his early *God's Grace and Man's Hope* (1949).

In this work as well as numerous subsequent articles, Williams sought to develop a liberal theology that was a deliberate alternative to neo-orthodoxy but one that nevertheless took neo-orthodoxy seriously. In much of this endeavor, Reinhold Niebuhr was Williams' conversation partner. One of his criticisms of Niebuhr centered on the latter's treatment of mutual reciprocal love. While fully aware that egoism is a part of any human relationship, Williams, unlike Niebuhr (who would make some shifts in this regard in the mid-1960s), saw self-affirmation and self-love (in the sense of self-worth) as indispensable to living and to any relationship. He had no use for the neo-orthodox bifurcation of the realms of science and the realms of faith. Using Whiteheadian process thought, the liberal alternative to neo-orthodoxy that he sought to carve out was always engaged in dialogue with and appropriating the insights of modern science, especially evolution and quantum physics, in the manner of his Liberal Protestant predecessors.

In Anglicanism, there were similar developments. Although neo-orthodoxy had an impact, in my view this impact was not as strong as it had been in Protestant (and later Roman Catholic) circles and the culture at large (at least with Niebuhr) in both continental Europe and the United States. Lionel Thornton (1884–1961) appropriated the insights of Whiteheadian process thought and sought to synthesize it with a more supernaturalistic understanding of God. Two other outstanding Anglican figures during this era were William Temple, who went on to become Archbishop of Canterbury, and Charles E. Raven.

Temple (1881–1944) is often treated as part of the reign of neo-orthodoxy. While a contemporary of the giants of neo-orthodoxy, Temple, as a good Anglican, did not believe in compartmentalizing anything and upheld the tradition of using all forms of human endeavor to worship God in "the beauty of holiness." He was a Christian Socialist who supported the Labor Party.

For Temple, God is One (in Three Persons) knowledge of whom cannot be bifurcated. Thus, he synthesizes faith and reason, scientific and other forms of knowledge. He uses Whiteheadian process thought to explain this and to argue for a contemporary version of nature being one of "two books" revelatory of the divine self. A chapter entitled "The Sacramental Universe" in *Nature, Man, and God* (1934) provides the best example of all of this, providing a synthesis of modern science, process thought, and the incarnational-sacramental ethos of Anglicanism. God expresses the divine self in the very materiality of the physical world and, just as the bread and the wine are transformed into the Body and Blood of Christ in the Eucharist, so is God ever continuously transforming, co-creating the universe that reflects the divine activity.

Charles E. Raven (1885–1964) anticipated some contemporary trends by advocating the convergence and integration of science and religion. In spite of Anglicanism's traditional openness to modern science, he felt the church needed to reinterpret the inherited tradition as it appropriated the insights of modern science, evolution in particular, if the Christian faith was to speak in an intelligible and meaningful way to contemporary men and women. In this doing so, Raven felt the church needed be self-conscious and intentional.

Raven's first academic endeavors dealt with the history of Christian Socialism. He was also a lifelong pacifist. In the early 1950s, he received a peace prize in Moscow. Raven was consistently and scathingly critical of the bellicose tendencies of all sides.

A scientist as well as a theologian, all of Raven's later theological works are attempts at the integration of religion and science. Most outstanding among these are *Evolution and the Christian Concept of God* (1936), *Jesus and the Gospel of Love* (1931). *Science, Religion, and the Future* (1994), and his two volume Gifford Lectures, *Natural Religion and Christian Theology* (1953). In Volume I of the Gifford Lectures, *Experience and Interpretation*, Raven argues that the Church Fathers of the first five or six centuries of the Common Era not only used the best science of their day but anticipated the theory of evolution. Volume II is an illustration how today we might do something similar, reinterpreting the faith completely in light of Darwinism.

Raven's last book was about the Roman Catholic paleontologist-theologian Pierre Teilhard de Chardin (1881–1955). Raven was enthusiastic not only about Teilhard's theology but about that kind of theologizing providing a pioneering model for the future in the relation between religion and science.

While Wieman took Chicago style liberal theology in a different direction and Meland was a bridge figure between two Chicago schools of liberal theology, after World War II there was the development of what we might consider bridge figures between neo- orthodoxy and what was to follow. The prototype of such a bridge figure is Paul Tillich (1886–1965). Although often considered within the neo-orthodox tradition, he was so sensitive to the cultural issues of his day and to theology wrestling adequately with those issues that it is more accurate to treat him as a bridge figure.

For our purposes, the most relevant of aspect of Tillich's theology is his "method of correlation." According to the method of correlation, culture, the broad range of human endeavors like literature, music, painting, sculpture, science describe the existential "the situation" of a particular era. Theology then needs to correlate the gospel, the good news that overcomes existential estrangement, with "the situation." The variety of human activities we see and hear in science, music, literature, painting sculpture, etc., shows us that the existential situation of the modern age is that of emptiness and meaninglessness. The task of theology is to correlate the good news that we are accepted as we are with the depth of our sense of emptiness and meaninglessness, for the good news overcomes those and all other forms of estrangement.

Although one could write at great length about Tillich, suffice it to say that his method of correlation opens a door for the science and religion dialogue. For Tillich, the relationship between religion and culture, religion and the situation, is a dialectical one, which can be a form of dialogue. He also embraces contemporary culture on its own terms, showing us the depths of our estrangement yet with each one of its many dimensions also acting as a "medium" of revelation.

Tillich was the first university professor fired by the new Nazi regime in 1933 and came to the United States as an immigrant that same year. His popularity and influence hit its peak in the 1950s and early 1960s. The historical context was considerably different from that immediately following World War I or the Great Depression. The end of World War II was greeted with euphoria and, in spite of the advent of Cold War, at least in the United States, there was considerable hope. Although there was some anxiety (because the Soviet Union shot its Sputnik into outer space before the United States did so, and because the Soviet Union sent Yuri Gagarin into outer space before the United States sent Alan Shepherd), the late 1950s and 1960s were characterized by optimism. Some of this was demonstrated symbolically by the election of the youngest president in American history; his opponent was only four years older. The Kennedy Administration rhetoric about a "New Frontier" and putting a man on the moon within ten years articulated this new optimism. Some intellectuals and journals pondered the "perfectibility of man."

In this historical and intellectual context, the neo-orthodox dominance collapsed as did the age of the theological giants—Barth, Brunner, Bultmann, the Niebuhrs, Tillich. What if anything, would take neo-orthodoxy's place? Who would replace the giants?

One eclectic group that in the context of the 1960s embraced its secular ethos was what I shall call "the theologians of secularization." The English words "secular," "secularization," "secularity," "secularism" are derived from the Latin word "saeculum," meaning "this world, this age." In one fashion or another all the theologians of secularization embraced radically "this world, this age."

A few definitions are in order (I am indebted to John Macquarrie for the distinctions; Macquarrie 1967). "Secularization" refers to the progressive removal of ideas and institutions from the dominance of religion. Originally used to describe the "laicization" of sacred places, the word came increasingly to refer to the separation of the religious and the secular, the sacred and the profane. For some, "secularization" became identical with "modernization."

"Secularism," as we have seen already in the theology of Karl Heim, refers to those thinkers who, in embracing the world, preclude any possibility of the existence of God. The world and its laws are perfectly capable of explaining the world and how it operates without any reference to God or nay sense of the sacred.

"Secularity," on the other hand, was a theological response that embraced the world yet was capable of affirming the existence of God. In fact, just as the theologians of "secularity" embraced the "worldliness" of the world, no less did they embrace the existence of God, insisting on the world's owing its very "worldliness" and ultimate significance to God.

Pivotal to the discussion was the figure of the German theologian-pastor Dietrich Bonhoeffer (1906–1945), who was imprisoned and executed by the Nazis for his involvement in the plot to assassinate Hitler during World War II. Bonhoeffer's ruminations during his imprisonment were published posthumously under the title *Letters and Papers from Prison*. Bonhoeffer was heavily influenced by Barth, including on the distinction between the Christian faith and revelation, yet also attempted to go beyond him. In his prison papers, he speculated that, since humanity had "come of age," it was time to develop a "religionless Christianity" and a secular way of speaking, preaching the Gospel. Although Bonhoeffer did not get a chance to elaborate on these ideas, they provided some fertile area of exploration on which all of the theologians of secularization would build.

The first volleys in response to this new situation were fired virtually simultaneously by some of the secularists and by some "moderate," advocates of secularity, although much of the conventional wisdom of the time considered them radical: Bishop John A. T. Robinson (1919–1983) in Great Britain and Harvey Cox (1929–) in the United States. Embracing modern science and its concomitant secularity in his *Honest to God* (1963), Robinson argued for a faith that was intelligible for contemporary human beings. He used Tillich, Bonhoeffer, and Bultmann among the theologians we have considered in this endeavor.

Harvey Cox, a professor at Harvard Divinity School, endorsed the process of secularization in *The Secular City* (1965) by claiming it was a part of God's work. God was the senior partner and humans were the junior partners in God's ongoing creation of the world. God calls humans to be in the world, embracing it, co-creating it, in the middle of secularization, of urbanization, and the cutting edge of technological innovation.

As early as 1961, Gabriel Vahanian (1927–) published *The Death of God*. Compared to some of the other "death of God" theologians, Vahanian was mild, referring to secularization as the disappearance of a unifying symbol system. For him, this was simply part of the "facts" of the contemporary situation. While other "death of God" theologians accepted this analysis, they wanted to go farther in carrying out the meaning of the term. Thus, Paul Van Buren (1924–1998), a former Barthian, in his *The Secular Meaning of the Gospel* (1963), maintained that given the evidence of science and the lack of empirical evidence for the existence of God, it is implausible for contemporary women and men to believe in the existence of God. He nevertheless believed that the pattern of call and response we see in Jesus Christ exists in all human beings in a new and radical way.

Beginning with his *The Gospel of Christian Atheism* (1966), Thomas J. J. Altizer (1927–) certainly went in a different reaction. Following Hegel and the second- century Christian heretic, Sabellius, who was accused of confusing and collapsing the three persons of the Trinity into one another, Altizer claimed that the Father had emptied the divine self completely into the Son, the perfect image of the Father. Thus, God really died on the cross and has been progressively manifest in the history of the human race, especially in human consciousness.

In a variety of ways and on a variety of grounds, the "death of God" theologians affirmed the possibilities of human existence and embraced most radically "the worldliness" of the world. But in order to affirm being human and embrace the world, theologians, in fact all humans, need to dispense with the idea of God.

Most of the death of God theologians were not particularly interested in the religion and science dialogue. They were interested in science only to the degree that science articulated the secularism of the age. Among them, William Hamilton took this secularism, eloquently described by Karl Heim, for granted, accepted it, and encouraged his fellow Christians to do likewise. Of all of the death of God theologians, Paul van Buren was the only one to wrestle with such an explicitly scientific term as "empirical" and to appropriate its narrow reference to experience as sense experience, the observable, rendering God "empirically unverifiable."

The theologians of "secularity" were not willing to accept the notion that one had to abandon all concepts in God in order to affirm the maturity and dignity of human beings and the worldliness of the world. One could believe in God in ways that did not make humans puppets at the end of a divine string or servile beings doing the bidding of a capricious deity. And in this endeavor, all affirmed and embraced modern science, most engaging in some sort of dialogue and/or attempting some convergence and integration.

Before I treat the theologians of secularity in some detail, we need to look at the changing historical context and ethos once again. The optimism of the early 1960s had been profoundly jarred. First, there was the shocking assassination of President John F. Kennedy. Secondly, in spite of the successes of the Civil Rights Movement, there was the persistence of the many faces of racism and the race riots of the middle and late 1960s. Additionally, many white liberal allies and supporters of the Civil Rights Movement had difficulty understanding the Black Power as well as other African-American movements that sought the empowerment of black people without the assistance of white allies. Finally, there was what to many was a loss of innocence, the war in Vietnam and the increasing military build up in Southeast Asia.

As I mention this profound change of sensibility by the mid-1960s, I need to point out that it did not change the theological landscape in terms of a radical embrace of secularization and of the world. If anything, embracing the worldliness of the world took on an urgency with the increased activism of clergy and laity.

John Macquarrie (1919–2007), who in the mid-1960s was in the process of switching from the Scottish Presbyterian Church to the Episcopal Church, made the distinctions between secularization, secularism, and secularity, to begin with (Macquarrie 1967). He grounded his work in as existential/ontological approach reminiscent of Tillich. In a manner quite congruent with the Anglican incarnational- sacramental ethos, he upheld God's affirmation of the world in creation, "the letting be" of the creatures, an interpretation of creation that affirms evolution; in the incarnation, the doctrine that God loves so much that she/he truly became incarnate as one of us; the Church, in which the incarnation continues; and in the sacraments, in which God expresses the divine self in the very physicality of the world. Macquarrie would affirm both God and God's radical affirmation of the world in this vein in subsequent writings, accepting modern science although not necessarily engaging directly in the dialogue.

Langdon Gilkey (1919–2004) was another notable theologian of secularity who affirmed both *God and the World*. The son of the liberal chaplain at Rockefeller Chapel at the University of Chicago, he grew up in the atmosphere of the glory days of the Chicago school, including during parts of the Shailer Mathews–Henry Nelson Wieman periods.

Gilkey's life defining experience came as a Japanese prisoner of war. He had gone to China to teach English and wound up spending the war years in a detention camp. He described his experiences in the moving *Shantung Compound* (1966). His war time experiences, combined with listening to Reinhold Niebuhr's lectures converted him to neo-orthodoxy. He earned his Ph.D. at Union Theological Seminary, studying with both Niebuhr and Tillich. Although Niebuhr remained the primary influence, Tillich's influence, as well as that of process thought, moved Gilkey beyond neo-orthodoxy to a more nuanced engagement with culture. After teaching at Vanderbilt University Divinity School, he spent most of his career at the University of Chicago as Shailer Mathews Professor of Theology.

In nearly all of his works, but especially in the ones that engaged science directly (Gilkey 1970; 1985), Gilkey saw science motivated by and pointing to ultimate questions about the meaning of life. While science was motivated by religious questions (we won't say that too loud!) and points to ultimate questions, science and religion do deal ultimately with different ways of knowing and dimensions of reality (this was very much

Figure 5.1 Union Theological Seminary in 1910 (Courtesy Library of Congress).

the gist of his testimony at the 1981 trial about Arkansas' "creationism" law which had mandated the teaching of "creation science" as a scientific alternative to Darwinian theory; Gilkey was the American Civil Liberties Union's star theological witness against the law) (Gilkey 1985). As mentioned earlier, while appreciative of his neo-orthodox background, Gilkey also transcended it, especially as he was increasingly influenced by process thought in his later years.

Beginning in the early 1960s, there was a "Science-and-Theology Discussion Group," that started to meet regularly. Their interest included the religion–science dialogue as well as the transformational impact of technological advances. Its members included William G. Pollard (1911–1989), physicist and Episcopal priest, who from the 1950s on was an enthusiastic advocate of dialogue, seeing the compatibility between chance and providence, to borrow from the title of one of his books. Another was Daniel Day Williams (1910–1973), who in a number of his works spoke of the "neo- naturalism" of process thought providing a helpful way of dealing with some of the thorny issues in the religion and science dialogue. By "neo-naturalism," he meant, first of all, that this world is all that there is and that God is in this world or nowhere. However, he also meant a world that was not mechanistic and deterministic but one in which the constitutive parts were interrelated yet dynamic and creative.

Even as the "Science-and-Theology Discussion Group" engaged science and religion in a positive way, the members of the group, along with nearly all of the theologians we shall deal with in the rest of this book, especially the process theologians, engage in a critique of neo-Darwinism, at times relying on quantum theory and more recently the discipline of ecology. Unlike Darwin himself, neo-Darwinist espouse a "sensationist (the only way we know things is through sense experience, as in our previous of British and scientific materialism)-materialist-atheistic" (reductionist materialistic) worldview that also, derivatively, maintain that the universe is meaningless and amoral (Griffin 2006, 7–17). As we shall see below, all of the theologians we treat reject this worldview even as they embrace evolution with sense of purpose yet also rejecting Intelligent Design. We need to keep in mind this critique of neo-Darwinism as we consider the more recent history of the Liberal Protestant embrace of science.

Within the context of the various responses to secularization, secularism, and secularity, as well as the activism of the 1960s and early 1970s, Ralph Wendell Burhoe founded the journal *Zygon*. Although Burhoe attended Harvard (1928–1932) and Andover Newton Theological Seminary (1935–1936), he never received a degree of any sort. Nevertheless, he became the Director of Harvard's Blue Hill Meteorological Observatory (1936–1946), and from 1947–1964 served as the first chief executive of the American Academy of Arts and Sciences. In 1964, he accepted a position as professor at Meadville/Lombard Theological School, the Unitarian Universalist seminary in Chicago.

In 1954, Burhoe founded the Institute of Religion in an Age of Science, which has played a vital role in the religion and science dialogue down to today. He was one of the founders of the Center for Advanced Study of Religion in 1972 and in 1988 of the Chicago Center for Religion and Science at Lutheran School of Theology.

Burhoe had a lifelong passionate concern for the relationship between religion and science. He thought that the profound sense of emptiness and meaninglessness of the contemporary age was due to the cleavage between science and religion. In his view, wholeness depended on the convergence, the "yoking" (the meaning of *Zygon*) of religion and science.

In Burhoe's fusion of science and religion, God is virtually identical with natural selection. However, natural selection was not synonymous with Social Drawinism. Rather, it was cooperation, trans-kin love that was the engine of evolution. He wanted to show that religion was a part of the evolutionary process. He argued that the human species was the repository of cultural evolution, which was then passed on to future generations.

Religion's vital role in cultural evolution was teaching love and especially trans- kin altruism. Correlated with this was a fascinating theory of immo–rtality: the stream of information that is one's personal center is released at the point of death into the larger stream of cosmic information and subject to an endless process of natural selection.

PROCESS THEOLOGY

Many of the proponents of the affirmation of both God and secularity were process theologians. Among the most important of the responses to secularization, secularism, and secularity among process theologians is that of Schubert Ogden (1929–). He maintains that in order for the Christian faith to be intelligible, it needs to affirm secularity. However, in order to be Christian, it also needed to affirm the reality of God. In order to affirm both God and secularity, Ogden seeks to integrate the insights of existentialism and process thought.

He maintains that to have some semblance of meaning, all human beings need to have a sense that they are contributing to something beyond themselves, beyond the moment. While seeking to contribute to one's future selves or to future generations in one's family or community provides meaning, it is not enough. All of these will perish in the flux of time. The ultimate significance of our lives can only be guaranteed if they contribute to something everlasting. That something everlasting is God.

Life is meaningful because we have some fundamental intuition that our lives matter. They matter because they are everlasting. And they are everlasting because they matter to the one who is everlasting. Thus, in spite of "perpetual perishing," the death of the immediacy and intensity of each moment, the meaning, the dignity, the intrinsic value of each moment are preserved everlastingly in the "consequent" nature of God, the receptive side of the divine life. The world and life in it are ultimately significant— precisely because the world and life in it matter to God in an ultimate and everlasting way!

Ogden is here clearly following Whitehead, appropriating the idea of the consequent nature of God, the receptive side of God in which God "feels" all "feelings" in the world. He is also following his own teacher, Charles Hartshorne (1897–2000). Hartshorne had been one of Whitehead's assistants at Harvard. One of the leading interpreters of his mentor, Hartshorne became one of the most original and innovative of philosophers during the long course of his life. He, of course, has a special place in the pantheon of process thinkers.

Hartshorne taught in the Philosophy Department at the University of Chicago from 1928 to 1955. Many of his students, including Ogden and John B. Cobb, Jr., whom we shall have occasion to mention below, were from Divinity School. He went on to teach at Emory University in Atlanta in 1955 and in 1962 was brought to the University of Texas by then Chair of the Philosophy Department, John Silber.

Hartshorne's presence at Chicago overlapped the tenures of Wieman, Meland, Loomer, and Williams. Unlike the others who stood within the radical empirical strand of the tradition, he developed the rationalist side of the Whiteheadian tradition. Hartshorne engaged scientists in dialogue and had some impact. He, like Whitehead, offers an organic, ecological vision of the world in which everything is interdependent yet creative. In other words, he, like Whitehead, offers an alternative to the mechanistic view of the world—an alternative that was supported by quantum physics and the new discipline of ecology and quantum physics.

It was in the context of the discussion of secularization, secularism, and secularity that pioneering work was done in the religion and science dialogue by three scientists who were also trained as theologians. The first one we shall consider, Ian G. Barbour (1923–), had a remarkable dual appointment in both the Physics and Religion Departments at Carleton College in Minnesota.With his *Issues in Science and Religion* (1965), he virtually created the science-religion dialogue in its current form. In that book as well as subsequent works that took into account ensuing developments in the scientific community, he engaged questions of methodology, the Big Bang theory, quantum physics, evolutionary biology, and genetics. He explored their theological implications.

One key issue, as we have already seen, is an organismic, ecological, holistic view of the world in place of the Newtonian-Cartesian mechanistic and deterministic one. Using his scientific background, Barbour illustrates that this new vision of the world arises from science itself, especially from the theory of evolution and quantum physics. He uses Whiteheadian process thought as a bridge in the dialogue.

As far as methodological issues are concerned, Barbour, both then and still today, argues for a radical empirical epistemology like the one we have discussed. In this epistemology, evident in James, Dewey, and Whitehead, the traditional bifurcation between object and subject, the traditional dichotomy between scientific objectivity and the supposed subjectivity of religious life is overcome: the objective world becomes a part of the subjective moment of experiencing as that moment feels the objective world; once the immediacy and intensity of the moment dies, that particular moment becomes part of the objective world that will be felt by the next moment of subjective experiencing.

Thus, in radical empiricism, the "observer" is involved in what is known, a "participant observer." Barbour draws on quantum physics and the works of such philosophers of science as Thomas Kuhn and Michael Polanyi to make the same point (Barbour 1997, 94–96). He espouses an epistemology (which is that of radical empiricism) that is called *critical realism*. According to critical realism, there is a real, objective world, independent of my experience. Yet, how that world is perceived also has an element of subjectivity. Thus, in answer to the question of whether or not a tree falling in the forest is making noise if there is no one to hear it, the answer of critical realism would be that, of course, the tree is making a noise; *when* the falling of the tree is heard, *how* it is heard is up to the subjectivity of the hearer.

For Barbour, the critical realist epistemology shared by modern science and process thought, quantum physics, with its emphasis on everything being energy; the Heisenberg uncertainty principle, which maintains that there is real indeterminacy in the universe; and an evolutionary, organismic, ecological biology that sees the world as becoming, dynamic, creative opens up new ways of science engaging religion in dialogue. Both science and religion using process thought as bridge, perhaps ultimately leading to integration and convergence.

Another great pioneer writing out of the same context of the 1960s, engaging the issues of secularization, secularism, and secularity, is the Australian biologist L. Charles Birch, one of the founders of the discipline of ecology. Birch is also a lay theologian . Synthesizing the works of such biologists as W.E. Agar and Sewall Wright, the process philosophy of Alfred North Whitehead and Charles Hartshorne, and his own work as a biologist, Birch argued in favor of a vision of the universe that was processive, ecological, dynamic, creative, holistic. In *Nature and God*, as well as numerous later works, he argued for the attribution of feeling, mentality, freedom, purposiveness in all subjective moments of experience, at no matter how rudimentary a level—from the tiniest energy event to amoebas to frogs to human beings (Birch 1965, 13–80). Like Barbour, Birch affirmed the God of process thought, luring the creatures to ever richer and more complex experiences, feeling their joys and sorrows, growing and changing with an ever changing, dynamic, creation.

The third scientist operating out of the context of the discussion of secularization, secularism, and secularity whom I shall consider is Harold K. Schilling, who was a Professor of Physics and Dean of the Graduate School at Pennsylvania State University. Like Barbour and Williams, a member of the previously mentioned Science-and- Theology Discussion Group, he was less well known than Barbour or Birch. Approaching the relationship between science and religion as a physicist and as a process thinker, his position was very similar to theirs, with a greater emphasis on mystery and on the future (Schilling 1973, 30–56, 257–276).

Having mentioned these three pioneers, I would be remiss not to mention the works of Richard H. Overman and Peter Hamilton. Overman

(1929–), a physician and Professor of Religion at the University of Puget Sound who had studied with John Cobb, sought a synthesis of evolutionary theory with process thought in his *Evolution and the Christian Doctrine of Creation* (1967). Hamilton, a mechanical engineer and Anglican priest, engaged evolutionary biology as well as contemporary scientific cosmologies in dialogue with process thought in his *The Living God and the Modern World* (1967).

Another process theologian who attempted to respond directly to the issues involved in the context of the religion and science dialogue was Kenneth Cauthen, who spent most of his career at Colgate Rochester Divinity School. Cauthen had already established his academic reputation with a very fine study of American liberal theology, *The Impact of American Religious Liberalism* (1962). In his *Science, Secularization, and God* (1969), in which he took the now familiar approach of affirming the secular, he used developments in evolutionary theory and quantum physics with process philosophy serving as a bridge to construct a new, ecological, organismic vision of reality. In this vision, there was room for a God who was luring the creatures to their fulfillment. Cauthen was quite eclectic in his approach using scientists, process thinkers, and some concepts that, while not directly influenced by, nevertheless paralleled some of those used by the Boston Personalists, who stood in the idealist tradition. The aspect of the Boston Personalism that he adopts is the notion that God's power is limited by a "dark impediment" (Cauthen 1969, 190). While this at times sounds like a cosmic creativity that can have ambiguous results, it is not very clear why reference to "a dark impediment" is necessary to explain the limitation of divine power.

Cauthen's interest in science also led to an intense interest in ecology, as one can see in his *Christian Bio-Politics* (1971) and *The Ethics of Enjoyment* (1975). He has been a prolific author. In such works as *Theological Biology* (1991) and *Toward a New Modernism* (1997), he has combined his early interest in the socio-historical method of the Chicago School with an abiding interest in contemporary science and in process thought.

The most prominent contemporary process theologian is John B. Cobb, Jr. (1925–). Born in Japan to missionary parents, he served in the military during World War II, translating captured Japanese documents. Following the war, he earned a masters and a doctorate at the University of Chicago. In the 1950s, after a brief stint circuit riding small churches and community college teaching, he went to Emory University. He accepted a professorship at Claremont School of Theology in 1959, a position that in due time became a joint appointment with Claremont Graduate School. An incredibly prolific author with a wide variety of interests, he established the Center for Process Studies in 1973.

For Cobb, it is simply irresponsible to do theology without engaging in dialogue with the sciences. In his *God and the World* (1969), written in response to the issues surrounding the problems of secularization, secularism, and secularity, he speaks of God as an "energy event" that *calls us forward.*

In 1969, under the influence of his son Cliff, Cobb had a virtual conversion experience in terms of the depth of his commitment to problems of eco-justice and sustainability. Subsequently, the religion-science dialogue, in a sense took on a greater urgency for him since it is integrally related to ecological issues.

In the early 1970s, Cobb was advocating a "new" Christianity committed to various forms of human liberation (from economic oppression, racism, sexism, anthropocentrism, and anti-Judaism), to experimental lifestyles for the sake of sustainability, and to rekindling a lost sense of community. The dialogue between and the integration of religion and science was a vital part of this "new" Christianity.

With the publication of his *Christ in a Pluralistic Age* (1975), Cobb abandoned the use of the language of a "new" Christianity and began using the idea of "creative transformation" instead. It is intriguing that while Cobb had been making very distinctive contributions to the speculative side of the process tradition, he was now reaching for a concept out of the empirical side, namely from Henry Nelson Wieman.

The model of creative transformation is dialogue—and vice versa. For example, if one of my students encounters new ideas in class, she may see them as threatening and consequently reject them. Or she may wrestle with the new ideas and let contradictory ideas simmer side by side. Or, in wrestling with those ideas, she may come to a new synthesis of her old and new that is in fact a novel, creative transformation of the old ideas as well as the new.

Cobb has used this model of creative transformation with a variety of issues. Prominent among these, given our world's religious pluralism, is inter-religious dialogue. For example, the dialogue with Buddhism, as is the case with all dialogues, needs to begin with an attitude of openness to the other, i.e., the hope that at the very least, we can learn from the other. Thus, what might Christians learn from the Buddhist idea of Emptiness, "Sunyata"? To Buddhists, emptiness is fullness and fullness is emptiness. This may sound quite contradictory and non-sensical. But to Buddhists emptiness is the result of not clinging to one's past selves, cherished beliefs, loved ones, possessions, or anything else. It means letting go of clinging and being the experience of the moment. And precisely because we have let go of clinging and have become the experience of the moment, precisely because we are empty, we are open to take more and more of the world into ourselves. Thus, emptiness is fullness and fullness is emptiness!

One thing that a Christian might learn and appropriate from Buddhists is to learn to let go and be the moment. One might, as Jay McDaniel has, use the Buddhist concept of Emptiness to make sense of the idea of the "consequent nature," the receptive side of God in which the divine self receives all experiences, taking more and more of the world into herself/himself (McDaniel 1985, 185–202). Buddhists, being non-theists, might very well object that this is still a form of clinging. We may very well have to live with such differences. But the dialogue has been enriching and creative transformation has occurred.

In a similar fashion, Cobb uses the model of creative transformation to engage modern science. Probably the best example of this is *The Liberation of Life* (1981), which he co-authored with Charles Birch. In that book, Birch and Cobb retell the story of evolution. However, they do so from the perspective of process thought, attributing feelings, mentality, freedom, intrinsic value to any momentary experience, from the tiniest energy event to ants to cats and dogs and to human beings. When it comes to God, they use the word "Life." Using both Wieman's distinction between the creative good and the created good and Whitehead's idea of God, they maintain that Life is the lure to greater complexity, to the experience of greater richness of experience, greater capacity for novelty as well as feeling all the feelings, human and non-human, in the world (Birch and Cobb 1990, 176–202).

In the years following his "conversion" on the issue of ecology, Cobb came to think of theology as critical thinking as Christians about issues of importance—such as eco-justice, sustainability, poverty, liberation in all its forms, interreligious dialogue, homosexuality, the right to die, animals rights, the American Empire, and much more. True to the best insights of Whiteheadian process thought, Cobb has sought to use its unifying vision to overcome the fragmentation of modern thought, particularly evident in the diffuse disciplines of the modern university. This collaboration with Charles Birch on *The Liberation of Life* and the manner in which they retold the story of evolution is a deliberate effort to overcome this fragmentation. Cobb has co-authored numerous books with people form other disciplines, most notably *For the Common Good*, co-authored with the economist Herman Daly. A book about sustainable economics, it challenges the prevailing economic orthodoxy about economic growth.

As I have been telling this story, I have slowly begun alluding to works written in the 1970s and 1980s. Needless to say, the socio-economic-cultural milieu was different and the issues changed. There was a retrenchment from the activism of the 1960s. Many of the activists of the 1960s, now older and with families to support, "joined the establishment." The activism of the 1960s was very much tempered by ethos of the 1970s. It was as though people had been collectively burned out on activism.

Nevertheless, the environment joined the causes that were important to people, particularly in the churches. Bioethics became a new subdiscipline as societies, especially in the West, were confronted with the fact that people were living longer and there were new technologies and drugs being discovered that could keep them alive even longer.

Perhaps, most importantly, there was a dramatic shift in the landscape: all the enthusiasm about secularization, secularism, and secularity seemed to be premature; there was a dramatic resurgence of religion throughout all of the world, a resurgence of the fundamentalist variety at that! The "counterculture" of the 1960 and early1970s already manifested an untraditional spirituality, religiosity, a search for transcendence. By 1970, Harvey Cox had written *Feast of Fools*, which, in contrast to *The Secular City*, sought an adequate ritualistic expression of the countercultural quest.

Far more novel was the increased involvement of evangelicals and fundamentalists in politics and all aspects other aspects of socio-economic-cultural life. Historically, at least the Scopes Trial in 1925, evangelical and fundamentalists were withdrawn from society; dancing was frowned on; involvement politics was considered improper and ungodly; televangelists were thought to have sold souls to the devil.

In the early 1970s, evangelical activism first showed its face on the left, toward the middle with Jimmy Carter, and finally with the elections of 1978 and then 1980 the Religious Right, comprised of fundamentalists of various stripes and their supporters, became pivotal players in American politics. Virtually simultaneously, there was the Iranian Revolution in 1979 and the rapid growth of "fundamentalist" Islam (to be sure, Wahabbi fundamentalism had dominated Saudi Arabia since at least the 1930s). In India, fundamentalist Hinduism was experiencing a rebirth and "fundamentalist" Jewish groups arose in the United States and Israel. In 1984, Harvey Cox wrote *Religion in the Secular City: Toward a Postmodern Theology.*

As we have seen in our discussion of Langdon Gilkey, under the influence of politically active, well organized and well funded, fundamentalists began pushing legislation that required, in some fashion, a presentation of the literal reading of the creation account(s) of Genesis as science in science classrooms. At first, as we have seen, this was done in states like Louisiana by requiring the teaching of "creation science," with its claims that there is scientific evidence for a six-day creation. Since the mid and late 1990s, the focus has been on "Intelligent Design," which we shall discuss below. While, as we shall see, there are some complex issues involved in the concept of "Intelligent Design," it is hard to deny that it has been used as a political issue very effectively by the Religious Right in states like Kansas and communities like Dover, Pennsylvania.

THE INTEGRATION OF SCIENCE AND RELIGION

It seems rather ironic that just as there has been a resurgence of religious fundamentalism there has been a growing movement in Liberal Protestantism whose members, like their predecessors, embraced the integration and convergence of science and religion. Headlines about fundamentalists and controversies involving religion and science dominate the news to such an extent that this Liberal Protestant trend, in fact Liberal Protestantism in general, is ignored. Yet, the spread of the movement for the integration of religion and science has been of sufficient strength to be described as the "religion and science community."

Some of the story we could have told earlier, for it goes back to Ralph Wendell Burhoe's founding of the journal *Zygon*, a pioneer in the dialogue and umbrella for a variety of voices in the field, including Williams, Barbour, and a number of scientists interested in the dialogue. Some younger theologians began to associate with Burhoe and *Zygon*. Indicative of the times, the religion and science dialogue started off as a hobby for most of them. Gradually, it came to define their professional careers.

Two of these theologians were Philip Hefner and Karl Peters. Hefner earned a Ph.D. at the University of Chicago Divinity School in 1962, and, after brief stints at Wittenberg University in Ohio and Gettysburg Theological Seminary, accepted a professorship at Lutheran Theological Seminary in Chicago in 1967. He retired from there in 2001. By the time Hefner returned to Chicago, he was deeply committed to reinterpreting the Christian faith, especially his cherished Lutheran tradition, in the light of contemporary science.

This reinterpretation, for Hefner, meant moving beyond dialogue to actual integration of the findings of science and theological interpretation. In this endeavor, Hefner was truly pioneering. He sought to integrate every field of science into the reconstruction of theological doctrines. He was one of the first theologians to take seriously and respond to the work of the sociobiologists.

A prolific author of books and numerous articles, the culmination of Hefner's work was *The Human Factor: Evolution, Culture, and Religion*. The central idea of the book, developed in numerous previous writings, is that human beings are *created co-creators*. Humans are products of a long creative process of biological and cultural evolution, thus "created." Here Hefner is using the sociobiologists' idea that while there is a relationship between biological and cultural evolution, the two are distinct. For example, sociobiologists typically maintain that biological evolution is characterized by the dominance of "the selfish gene," the love of one's kin and not beyond. It is only in cultural evolution that we learn altruism, love of the other, here in the sense of love beyond our own kin. Culture (and religion, although sociobiologists disagree among themselves on this point; Richard Dawkins, for example, is a militant atheist and vehemently anti-religious while E.O. Wilson sees some pragmatic uses for religion) plays a vital role in the nurturing of altruism.

Hefner, however, is not a determinist. Though we are shaped by all sorts of environmental factors, genetic influences, etc., we are free and responsible for what we do with those factors and influences and the range of opportunities open to us. He writes, referring to the word "created":

First, it recognizes that the human being is placed within an ecosystem, in an intimate inter-relationship with an environment that conditions human beings in significant ways. *Created* also refers to the genetic component of the two-natured creature, underscoring that neither the individual nor the group has control over the inheritance of our genetic composition, our genotype... The genetic component is also characterized, however, by the processes that allows the free human creature to emerge. Cultural conditioning is incorporated into the term *created,* although it must be noted that culture is the most dramatic locus of the human being's freedom. The genetic and cultural share in both conditionedness and freedom. The genotype is the source of our being conditioned, but it also provides the resources for freedom to emerge, else we would not be free. Conversely, although freedom reaches its most efficacious form in human beings, that culture is both conditioned by genes and environment and also is a basis for conditioning traditions that shape each new generation. (Hefner 1993, 36)

Furthermore, he maintains that "the image becomes genuinely theological when we include the assertion that humans have been created and given their place in the process by God" (Hefner 1993, 36). He maintains that:

In its theological dimension, therefore, the first element in this summary concept speaks of the primacy of God and the divine creating activity. Whatever we mean by the term *God* and whatever conceptuality we employ for thinking about God, that God is the ground of this process in which humans have emerged. (Hefner 1993, 36)

Co-creator is synonymous with freedom, which "... refers to the condition of existence that in which humans unavoidably face the necessity of making choices and of constructing the stories that contextualize and hence justify those choices" (Hefner 1993, 38). With that freedom comes a radical responsibility to God, for ourselves, for the creative processes of nature that produced us, and especially for the numerous technological innovations that make our lives simpler and help us to live longer but also threaten life on the planet.

Hefner's concept of the *created co-creator* is reminiscent of both process thought and the thought of Reinhold Niebuhr. In process thought, a momentary experience arises out of the context of myriads previous such moments, out of its own past, that of its immediate environment, of the universe itself. But the momentary self is free as to what it does with that past, *how* it appropriates that past as well as how it responds to the possibilities of the future. In similar fashion, Niebuhr maintains that on the one hand, we are limited by our finitude—and this certainly included limitations we have acquired genetically. On the other hand, we are free: we have the capacity, in however small a way, to transcend those limitations. This observation is not meant to detract in any way from the originality of Hefner's proposal.

Hefner's good friend Karl E. Peters collaborated with him on many projects, particularly the yearly Star Island conferences off the coast of New Hampshire, sponsored by the Institute on Religion in an Age of Science (IRAS) and *Zygon*. Peters finished his Ph.D. at Columbia University on Henry Nelson Wieman in 1966 under the direction of Daniel Day Williams. The influence of both Wieman and Williams would be lasting. Peters accepted a professorship at Rollins College in Florida and stayed there until his retirement in 2001.

No less than Hefner, Peters sought the integration of science and religion in theological construction. Like Hefner, he engaged all of the sciences—astronomy, evolutionary biology, sociobiology, etc. Through their involvement with IRAS and *Zygon*, both Hefner and Peters, a Unitarian Univeralist, became friends of and collaborators with Ralph Wendell Burhoe. Their respective attempts to integrate science and religion were very much influenced by Burhoe. Peters was the long time editor of *Zygon* (he and Hefner are coeditors).

Some of Peters' most powerful pieces of writing are on death. In a 1987 editorial in *Zygon*, he very eloquently and movingly wrote about the death of a friend. In a compelling synthesis of Darwinism, chaos theory from physics, and aspects of Hinduism (the idea that destruction precedes creation), he argues for seeing death as meaningful because it is part of the rhythm of nature, of the ebb and flow of time in which the old must die so that the new might live. Nine years later, he would say something similar upon the death of his first wife, Carol.

Decades before, Charles Birch had written in a similar vein about the "cruciform" pattern of the universe, that is to say, the death of the old in order that there may a newness of life. Peters, reflecting existentially on his own personal experience and combining it with new developments in the sciences (chaos theory) and interreligious dialogue, gave an articulation of this sensibility with new power.

As we can see from this discussion, Peters, like Wieman before him, sought to develop a consistently naturalist theology. The culmination of this endeavor, and of all his work, is his *Dancing with Sacred: Evolution, Ecology, and God* (2002). There is not only no God beyond or above the world of nature, neither is there a personal deity, although it is understandable why many people find a personal God more accessible.

The meaning of existence can be found in "dancing with God." Peters approaches this by analyzing the role of "leading" in dancing:

> The best kind of dancing is when no one leads, when the leading is back and forth sharing, when each party responds to the subtle movements, touches, gestures, and words of the other. When this happens both parties give themselves fully to the dance of dynamic relating. Then the relationship becomes a beautifully flowing movement of two people interacting with one another. Over time this can create beautiful patterns of creative friendship, partnership, and marriage. The key to this kind of zestful living is that neither participant is trying to advance his or her private goals. There is, in fact, no goal except the dance itself, being together in living life. (Peters 2002, 46)

Peters' image of evolution and of nature is that of dancing, "dancing with no one leading but with all participating and mutually influencing one another" (Peters 2002, 47). God is the music and responding to it "...brings one into relation with our sacred center" (Peters 2002, 49).

Thus, dancing is just for the sake of dancing. The "payoff" in this is greater openness to others and to the non-human natural world. But the greatest payoff, according to Peters, is "...participating fully in every moment of life" (Peters 2002, 50). While we may have goals and purposes that are important, "if we are not open to our goals and ideals becoming transformed by the grace of the dance, we may miss out on the joy of being in relationship with the divine in our midst" (Peters 2002, 50). Thus, he concludes that:

> In learning to dance with the natural world around us and with other human beings, we become more alive. This is the big payoff. We become more in tune with ourselves, others, and the natural world. We see more, experience more, enjoy more. We become part of the dance of the sacred—the dance of that system of interactions in the universe and society that brought us into being, that sustains us in our living, and that continually transforms as part of the ever changing future. (Peters 2002, 51)

Robert John Russell (1946–) is a theologian-physicist who has used what he calls "the creative mutual interaction" model to come very close to the integration and convergence model in his own work (Peters and Hewlett 2003, 150). An ordained minister in the United Church of Christ, he has an M.S. in physics, a B.D. and an M.A. in theology and a Ph.D. in physics. With his first teaching position at Carleton College (1978–1981), he has been very much influenced by Ian G. Barbour, who was on the Carlton faculty.

Since 1981, he has been teaching science and theology at the Graduate Theological Union, Berkeley, California. The same year he founded and became the director of the Center for Theology and the Natural Sciences (CTNS) in Berkeley. Under Russell's leadership, CTNS has been at the cutting edge of the science and religion dialogue, granting degrees and providing introductory and advanced courses for seminary and graduate students in the area. CTNS has also developed and helped pastors develop courses in the science–theology dialogue.

Russell has been the editor of a number of volumes on the dialogue, some of which have been published under the auspices of the Vatican Observatory and University of Notre Dame Press, others under the auspices of CTNS and the Vatican Observatory. Of these volumes, containing the papers from yearly conferences at the Vatican Observatory, probably the most famous are *Physics, Philosophy, and Theology* (1988), *John Paul II on Religion and Science* (1990), and *Quantum Cosmology and the Laws of Nature* (1993).

Russell's main concern is the problem of divine action. He upholds the Heisenberg uncertainty principle as basic to the freedom of humans and non-humans. Russell locates divine action in the moment just prior to the indeterminate becoming determinate, a process in which God and nature interact. Russell calls this "noninterventionist objective divine action" which means that although God acts, she/he does not suspend the laws of nature, in contrast, to the modern usage of the word "miracle" suggests. The more basic the unit of reality, the more God is active and involved, while in more complex creatures with greater capacities for freedom, God is less active (Peters and Hewlett 2003, 155–158).

Ted Peters (1941–) is another very important figure in the religion–science dialogue and the contemporary move toward their integration and convergence. Peters was yet another University of Chicago Ph.D. An ordained minister in the Evangelical Lutheran Church of America, he served several pastorates. He has been at Pacific Lutheran Theological Seminary in Berkeley, California since 1978.

Peters initial interest was on science as a cultural phenomenon and, influenced by Langdon Gilkey, the ways in which science points toward transcendence. He was also influenced by the future orientation of the German Lutheran theologian Wolfhart Pannenberg as well as his attempts to reestablish the unity of all knowledge in light of the growing fragmentation of the disciplines. Pannenberg has been a consistent Hegelian who has tried to "historicize" nature, i.e., stress its contingency and capacity for creativity and novelty, endeavors that have influenced Peters.

(Having mentioned Pannenberg, some readers may be surprised that I make no allusion to Jürgen Moltmann, who is usually mentioned with Pannenberg as a future oriented theologian of hope, and who in a survey done by the Liberal Protestant journal *The Christian Century* in 1975 was considered worldwide to be the most influential Protestant theologian since the middle of the twentieth century. I have not examined his thought in this book because, although he has written extensively on ecotheology, it was independent of the religion–science dialogue that was so important to the historical period I am describing and is pertinent to the theme of the book).

Soon after his move to Berkeley, Peters came into contact with Robert John Russell. It was the beginning of a fruitful professional and personal relationship. He has been the director of the CTNS Science and Religion Course Program. Peters has been a prolific writer, the author of *The Cosmic Self* (1990), *God—The World's Future* (1992, 2000), *God as Trinity* (1993), *Sin* (1994), and *Science, Theology, and Ethics* (2003). He edited *Cosmos as Creation* (1989), *Science and Theology: the New Consonance* (1998), and *Bridging Science and Religion* (2002). He is also the editor of the journal *Dialog: Journal of Theology* and co-editor of *Religion and Science*. In all of his works, Peters has appropriated the insights of the sciences in his constructive work as a systematic theologian, stressing, following Pannenberg, the indeterminacy and openness of the future.

Much of Peters' interest has focused on ethical issues, genetics in particular. His most important book *Playing God? Genetic Determinism and Human Freedom* (1997, 2002), now in its second edition, illustrates the centrality the issue of genetics has for him.

One aspect of his concern with ethical issues surrounding genetics has been to reject genetic determinism and to affirm the reality of freedom. While, to be sure, we cannot change our genetic inheritance, there are always choices available as to what we can do within a narrow range of possibilities. This kind of affirmation of freedom is fundamental to the plausibility of ethics.

Secondly, Peters' perhaps surprising advocacy of genetic engineering hinges and builds on freedom. Like most process thinkers (and it is not inaccurate to call Peters a process thinker, although he is not a Whiteheadian), he maintains that human existence is so plastic, so dynamic that there is no such thing as "human nature." To be sure, as mentioned previously, our genetic inheritances are given but that does not mean we have a static human nature. We have freedom and radical responsibility as to what we do with our possibilities within a range of probabilities. If this is so, then, Peters, argues, human existence is so fluid, dynamic, creative that we need to be open to genetic engineering, cloning, stem cell research. Engaging responsibly in such endeavors is truer to being human than trying to stop such endeavors in order to safeguard a non-existing static human nature. His argument also rests on pragmatic grounds: if genetic engineering can lead to discoveries that will cure disease and alleviate pain, it is intrinsically worth doing.

At this point, I would like to turn to two Anglican theologian-scientists, Arthur R. Peacocke and John Polkinghorne. Peacocke (1924–2006) was a physical biochemist who pioneered research into the physical chemistry of DNA early in his career. Always searching and questioning yet committed to the church, he was ordained a priest in 1971. He has had a distinguished career as a priest-theologian-scientist.

In 1971, Peacocke published *Science and the Christian Experiment*, for which he won the LeCompte du Nouy prize. He was also the author of *Creation and the World and the Science* (1979), *Intimations of Reality* (1984), *God and the New Biology* (1986), *Theology for a Scientific Age* (1990, 1993), *Paths from Science to God* (2001), and *Evolution: The Disguised Friend of Faith?* (2004). In 1973, he became Dean of Clare College, Cambridge. His retirement was a busy one: he was Warden Emeritus at the Society of Ordained Scientists; Honorary Canon at Christ Church Cathedral, Oxford; Vice President of the Science and Religion Forum of Modern Church People's Union; and Council Member of the European Society for the study of Science and Theology.

A pervasive theme for Peacocke is a Aubrey Moore's claim that evolution is "the disguised friend of faith" (Peacocke 2004). For Peacocke, evolution works, from the Big Bang onwards, through natural selection. This involves the interplay of law and chance that pervades the universe and is its foundation. While "things make themselves," God is also immanent in the creative processes of nature, in the interaction of law and chance. God prods toward the realization of what exists as possibility.

Peacocke uses the powerful image of the composer to describe God's role in creation:

Or...is more like a composer, who, beginning with an arrangement of notes in a simple tune, elaborates and expands it into a fugue by a variety of devices of fragmentation and reassociation; of turning it upside and back to front; by overlapping these and other variations of it in a range of tonalities; by a profusion of patterns of sequences in time, with always the consequent interplay of sound flowing in an orderly way from the chosen initiating ploy (Peacocke 1979, 2004, 105)

He elaborates:

In this kind of way might the Creator be imagined to unfold the potentialities of the universe he himself has given it. He appears to do this by a process in which the creative possibilities, inherent by his own intention, within the fundamental entities of that universe and their interrelations, become actualized within a temporal development shaped and determined by those self-same inherent potentialities he conceived from the first note. (Peacocke 1979, 2004, 106)

There is more to the image of the composer. The composition is unfinished and is still being written as it is being played. The composer as well as each musician is making adjustments as they are in effect co-composing, co-creating the musical piece. So it is with God creating through the evolutionary process, with each entity responding to God and to all other entities, and God responding to the response of each entity, in a web of relationships.

Using the image of the dance of Shiva from Hinduism, Peacocke also discusses the notion of the dance of creation. The dance of Shiva is the dance of creation, preservation, and destruction. Peacocke appropriates the image to highlight the interplay of law and chance, God's provision of order in the universe as well as the divine exuberance and play in taking the risk of giving the divine self to the universe, of encouraging zest and novelty (Peacocke 1979, 2004, 106–111).

The creative evolutionary process drives at greater and greater of complexity, greater degrees of complexity in the pattern of self-organization. The greater the degree of complexity, the greater the capacity for freedom and novelty. Creation through evolution is continuing creation or creation as continuous, *creatio continua*, a favorite theme throughout the history of Liberal Protestantism that we have told.

Peacocke is a free will theist, that is to say he upholds the notion that God freely decides to limit divine power in order that the creatures may be able to exercise their freedom, that they may have the capacity to return God's love freely. Just as death is part of natural process, so is pain and evil in a world in which freedom is real. Sin, pain, and evil are transformed by a God who participates in the lives of her/his creatures. Peacocke maintains that "*God suffers in, with, and under the creative process of the world* with their costly, open-ended unfolding in time" (Peacocke 1993, 126). This is what the incarnation and cross of Jesus Christ are about. Thus, although not a process theologian, he is a panentheist, adhering to the notion that God is in everything and everything is in God.

Peacocke works out of the incarnational-sacramental ethos of Anglicanism. The incarnation is the ultimate affirmation of the world, God's love being such as to become a human being. The sacraments are no less world affirming. Peacocke appropriates Temple's concept of a sacramental universe, that God expresses the divine self through the very physicality of the universe. Thus, in the sacraments, creation and the new creation come together, as does the transformation of what has been given us through human work (Peacocke 1971, 187–188).

In closing this part on Peacocke, I would like to note that, unlike the process theologians and scientists, Hefner, Karl Peters, Russell, and Ted Peters, who situate humans squarely in the non-human natural world, the Anglican theologian-scientist remains a dualist when it comes to the relation between human and the non-human natural world. That is to say, he is anthropocentric, human-centered, with humans being superior to the non-human natural world while the others attempt to liberate the Christian tradition from anthropocentrism.

The other Anglican priest-theologian-scientist I would like to mention is John Polkinghorne (1930–). He was a professor of theoretical physics for twenty-five years, a professor of mathematical physics at Cambridge (1968–1979), and a Fellow of the Royal Society (1974). He resigned his professorship to study for the Anglican priesthood and was ordained a priest in 1982. After a few years in the parish ministry, he became Dean of the Chapel, Trinity Hall (1986–1989). He was president of Queens' College, Cambridge until his retirement in 1996. Polkinghorne is yet another prolific author. Among his best known works are *Science and Providence*, *The Faith of a Physicist* (1994), *Belief in God in an Age of Science* (1998) and *Science and the Trinity* (2004).

Polkinghorne argues that if there is one God, then there is a unity to all knowledge, including scientific and religious knowledge. Theology is the discipline that can unify all the various branches of knowledge, a notion resisted by many scientists. Polkinghorne, as a physicist, makes use of quantum theory for the fundamental unity and interrelatedness of all things.

Like Peacocke, Polkinghorne adheres to the free will defense when it comes to the problem of evil and God's power. In other words, God voluntarily relinquishes some power in order that the creatures may have some exercise of freedom. He combines this free will defense with "kenotic" theory, "self-emptying." This is the idea that, because God loves the world, God pours out or "empties" the divine self into the creation, into Jesus Christ as the incarnation of God, and into the life of the Church as the continuation of the incarnation. In the context of his use of the free will defense and kenotic theory, he makes use of quantum theory, Heisenberg's uncertainty principle, and chaos theory to establish freedom in the universe.

Yet another theologian who typifies the integration and convergence of religion and science is Philip Clayton (1956–). He is Ingraham Professor of Theology at the Claremont School of Theology and taught previously at California State University, Sonoma. He has been a visiting professor at Harvard Divinity School and at the University of Munich. He is the author of *From Physics to Theology* (1989), *God and Contemporary Science* (1997), *The Problem of God in Modern Thought* (2000), *Mind and Emergence* (2004), and co-editor of *In Whom We Live and Move and Have Our Being* (2004) and *Evolution and Ethics* (2004) among others. He is also the author of numerous articles.

Clayton's work has engaged physics. More recently, he has turned his attention to emergence theory. Clayton's point of entry into the religion and science dialogue is the idea that science by itself is incapable of providing an adequate view of the universe. Alluding to contemporary scientific discussions of cosmology, he claims that "... *the single greatest positive result of current discussions in cosmology lies in the fact that scientific results plead for meta-physical, and ultimately theological, treatment and interpretation*" (Clayton 1997, 160–161). Engaging physics, scientific cosmology, and emergence theory, Clayton develops a process metaphysics that is the bridge to a panentheistic doctrine of God. While quite congenial with Whiteheadian process thought, his process metaphysics and panentheism is not explicitly Whiteheadian and draws from a variety of sources.

Clayton sees emergence theory as providing a particularly fruitful focal point for the science–religion, science–theology dialogue. Emergence theory deals with finding the laws that govern the universe's tendency for greater and greater complexity in self- organization. These laws are probabilistic and not mechanistic. In this world, there is real randomness, chance, and freedom. God and God's creatures jointly compose the melodies of the cosmos without knowing the final outcome (Clayton 2004). Clayton alludes to Philip Hefner's notion of "created co-creators" as an eloquent description of our identity as human beings (Clayton 2004). God also genuinely feels the joys and sorrows of her/his creatures, truly and eminently participates in their lives.

I would like to return at this point to Whiteheadian process thought and how it has been used to seek the integration and convergence of religion and science by such process thinkers as David Ray Griffin, John F. Haught, and such younger scholars as Jay McDaniel and Nancy R. Howell. We have already considered efforts at integration and convergence by such process theologians and/or scientists as Ian G. Barbour, Charles Birch, Kenneth Cauthen, and John B. Cobb, Jr. Now we are considering younger theologians, who, if anything, take the integration and convergence of religion and science more as a matter of course, unencumbered by the shadow of neo-orthodoxy.

David Ray Griffin (1939–) studied with Cobb at Claremont Graduate School, now Claremont Graduate University, then went on to teach at the University of Dayton (1968–1973). He returned to Claremont in 1973, becoming director, with John Cobb, of the newly established Center for Process Studies. Griffin has written and edited books too numerous to mention here. Among the more important are *A Process Christology* (1973); *God, Power, and Evil* (1976); *Process Theology: An Introductory Exposition* (1976) with Cobb); *Physics and the Ultimate Significance of Time* (1985); *God and Religion in a Postmodern World* (1989); *Evil Revisited* (1991); *Parapsychology, Philosophy, and Spirituality* (1997); *Unsnarling the World-Knot* (1998). Particularly significant for the religion and science dialogue are *Reenchantment without Supernaturalism* (2001); *Religion and Scientific Naturalism* (2000); and *Two Great Truths* (2004). Griffin was the editor of the SUNY Series on Constructive Postmodern Thought and Director of the Center for Postmodern Thought in Santa Barbara, California.

In the 1980s, Griffin engaged in dialogue with the works of such physicists as David Bohm and Ilya Prigogine that were quite congenial to the relational, ecological view of the world presented by process thought. Also congenial to process thought is their view of the world as dynamic and creative, with the future being indeterminate (Griffin 1985). Griffin's concern was to develop a "postmodern worldview," which was ecological and relational, in which freedom and creativity are real in contrast to the mechanistic and deterministic view of much of modern science. He sought the "reenchantment of science."

In his more recent works that deal with religion and science, Griffin has sought to extend this process-relational vision. He has taken on scientific materialism and has proposed, in an argument reminiscent of Wieman and Williams yet far more elaborate and highly developed, process thought as an alternative form of naturalism in which God's existence can be quite plausible even to the scientific mindset. He makes use of the radical empiricism of Whitehead, with its defense of the primacy of non- sensationist perception as a way to build bridges to scientific knowledge. The God of process theism is not a God above or beyond nature but in it. Thus, he articulates eloquently how the process doctrine of God is a form of "evolutionary theism," "naturalistic theism."

Griffin makes it very clear his view is not pantheistic, that is to say, equating with the totality of the world or the web of life as Loomer did late in his career. It is very clear that his position is panentheistic: God is in the world and the world is in God.

One of the most important contributions Griffin makes to the theological side of the discussion is his rejection of the doctrine of "*creatio ex nihilo.*" Historically, most theologians since the second century have espoused *creatio ex nihilo*, the creation of the world out of nothing. In other words, there was no world of any sort before God's decision to create world. Most theologians have thought this was extremely important to defend because in doing so they were defending the priority of God's action, of God's prevenient grace before any creaturely action.

Historically, *creatio ex nihilo*, creation out of nothing, emerged as a response to the Marcionite heresy. Marcion was a second-century theologian who maintained that the physical world was evil because of its physicality. If that was the case, its creator also had to be evil. Thus, in effect, there were two Gods: the evil God who created the evil physical world and Jesus Christ, the good God who delivers us from this world. *Creatio ex nihilo* was a handy way of condemning Marcionism, affirming that God is the creator of all things, including the physical world. And if God is all good, then, as the biblical witness affirms, the physical world is good, reflecting the goodness of its creator (Griffin 2004, 36–44).

The doctrine of creation out of nothing has, according to Griffin, floundered on a number of points. For one, it is hopelessly tied to supernaturalism. For another, in affirming that God created everything, *creatio ex nihilo* makes God responsible for evil, compromising God's goodness. In addition to being inseparable from the theodicy question (the problem of evil; if God is all good and all powerful, why is there evil in the world), it compromises the freedom of the creatures and of the world (Griffin 2004, 36–40).

Griffin argues instead for creation of order out of chaos—in a manner reminiscent of Plato. The physical world and God have always existed simultaneously—it is virtually impossible to think of a time when there was not some kind of world. God, in ordering possibilities in order of graded relevance, creates order out of chaos. With God luring all actualities to their self-realization in their fundamental interdependence with each other, always acting persuasively and not coercively, God and all of God's creatures *co-create* the world. Moreover, typical of process thought and contrary to the free will defense, God does not choose to limit divine power; God is already doing all that God can, being subject "to the rules of the game," being not an exception to metaphysical categories but their chief exemplification. God may be the greatest power but not the only power; anything that is has some degree of power. The creatures thus have the freedom to resist the divine grace (Griffin 2004, 36–58). To Griffin, creation as the creation of order out of chaos rather than out of nothing, supported by a number of contemporary biblical scholars, is more consistent and coherent as far as the issues we have mentioned above are concerned. Moreover, it is more adequate to the facts as depicted by modern science, an interdependent and interrelated world that is creative and in which indeterminacy is real.

The next theologian I shall consider is John F. Haught, whose typology of the relation between religion and science we introduced in Chapter 1. Although he is a Roman Catholic, I include him in this chapter because of the closeness of his thought to that of a number of Liberal Protestant theologians we are considering in this chapter; because his work is among the most important contemporary attempts at integrating science and religion; and his own unique insights using process thought. He is the author of *Nature and Purpose* (1980), *The Cosmic Adventure* (1984), *Who Is God ?* (1986), *The Revelation of God in History* (1988), *What Is Religion?* (1990), *The Promise of Nature* (1993), *Mystery and Promise* (1993), *Science and Religion* (1995), *God After Darwin* (2000), *Responses to 101 Questions about God and Evolution* (2001), and *Deeper Than Darwin* (2003). He is Professor of Theology at Georgetown University and Director of the Georgetown Center for the Study of Science and Religion.

Throughout his career, Haught has engaged the religion and science dialogue. He thinks it is a mistake for theologians to ignore science, evolution in particular in their constructive task (Haught 2000, 1–2). Instead, he maintains that "... Darwin has gifted us with an account of life whose depth, beauty, and pathos—when seen in the context of the larger cosmic epic of evolution—expose us afresh to the raw reality of the sacred and to a resoundingly meaningful universe." In a manner typical of process theologians and the other Liberal Protestant theologians we have been considering, he rejects the notion that scientific materialism is a metaphysical position that necessarily follows from Darwinism. Instead, he sees Darwinism as quite compatible with process thought's evolutionary theism and its vision of the world as dynamic, creative, constituted by interrelated parts (Haught 2000, 11–104). In effect, he argues for a synthesis of Darwinism and process thought.

Haught makes use of the idea of "information." He writes that "by 'information' I mean, in a broad and general sense, the overall ordering of entities—atoms, molecules, cells, genes, etc.—into intelligible forms of arrangements" (Haught 2000, 70). This is something beyond the scope and limitations of scientific materialism, yet discernment of patterns of order is important to scientists. This idea of information provides a useful bridge between science, philosophy, and theology (Haught 2000, 70–80). While there may be information of a more technical variety, "information" in Haught's sense of the word is something used by scientists and theologians alike.

Haught also makes use of the idea of "hierarchy." While he favors liberation from oppressive forms of hierarchicalism and domination, he like, Peacocke, sees nature replete with increasingly complex hierarchies of complexity and self organization, a kind of hierarchy vital to our experience and understanding of the sacred (Haught 2000, 57–80).

In Haught's Whiteheadian concept of God, she/he lures all actualities to realize themselves in their fundamental interdependence with all other entities. God, a "suffering God" in a "cruciform" universe, also shares and participates in the lives of his/her creatures. Haught is eloquent in making use of the idea of "kenosis," "emptying," referring to the love of God being such as to empty the divine self into creation and into Jesus Christ, especially as we see the self-emptying of God on the cross (Haught 2000, 104–120). He also uses the concept of the "divine letting be" of the world, God's co- creating the world by allowing it to be itself, with the freedom of its constituent parts (Haught 2000, 104–120).

As powerful and eloquent as his use of the ideas of kenosis and "letting be" are, the use of these ideas seems to be in tension with Haught's appropriation of the process understanding of God. For one, the language of "letting be" is suggestive of a divine decision to limit the exercise of divine power, in other words the free will defense. It remains unclear whether or not this was Haught's intent.

The use of the concept of kenosis is also in tension with a very basic tenet of process thought, namely the notion that there is a "self" to empty. While this gets into some technical issues in Whiteheadian process thought that are beyond the scope of this book, a brief mention of the issue involved seems appropriate.

Contrary to most of Western thought (and Hinduism), there is no "substantial self," an enduring self that does not change in spite of the changes it may undergo in appearance, like the core of the apple that remains even after the apple has been peeled. Rather, the self is a momentary self, the "event" of the moment (as we have seen before). Since God is the chief exemplification of metaphysical categories, this is true of God as well. Thus, the language of self-emptying is problematic and in tension with the process understanding of the self.

Of course, it is possible to understand the self in self-emptying as the momentary self as it receives data from its environment, as it becomes more and more open to and receives more and more of the world around it into itself. Haught's identification of kenosis with a suffering God, particularly as we see that suffering in the image of Christ on the cross, suggests that this is his understanding.

As I have just suggested with the idea of kenosis, in spite of the difficulty of the use of language, it is possible to reconcile both the ideas of self-emptying and "letting be" with the process understanding. Hopefully, Haught will provide more clarity on these issues.

A very original and creative aspect of Haught's work has been the construction of a worldview in which human beings belong to the universe. This is in contrast to the scientific materialism perhaps most powerfully expressed by such scientists as Jacques Monod (1910–1976) and Steven Weinberg (1933–). For Monod, humans are the product of chance as well as necessity, with no purpose or meaning to the universe or to human existence. Steven Weinberg, in his oft quoted statement, has written that "the more comprehensible the universe has become to modern science, the more it seems pointless" (Weinberg 1977).

The views of both Monod and Weinberg are reminiscent of the existentialist Albert Camus' *The Myth of Sisyphus*. In the ancient myth, Sisyphus is condemned by Zeus to roll a boulder up a hill only to have it roll back down as he approaches the top. The process is repeated endlessly. For Camus, Sisyphus is the paradigm for the absurdity of human existence. However, the dignity of human beings is visible in Sisyphus's determination. Weinberg seems to express a similar position, adding to his statement about the "pointlessness" that humans do create meaning in art, literature, science, personal relationships, etc.

Haught maintains that this kind of thinking is an expression of our profound alienation from the non-human natural world, an alienation which, to anticipate the next chapter, has led to the environmental crisis. As an alternative, he presents his Whiteheadian vision of a creative, participatory universe, in which all things interrelated, in which all things belong, in which all actualities seek the experience of beauty. And this is a vision to which Darwin made an immense contribution!

In his recent writings, Haught has taken on the advocates of Intelligent Design, the most articulate and famous among them being William Dembski and Michael Behe. Intelligent Design advocates argue that the Darwinian notion of natural selection is inadequate to explain the order and complexity of organisms; there is complex design in organisms that suggest a "designer" (Haught 2003, 87–89). Devotees of Intelligent Design claim that they make this move on scientific rather than metaphysical grounds (Haught 2003, 88).

In spite of protestations to the contrary by people like Dembski and Behe, Haught asks, "What else, though, could the notion of 'intelligent design' be pointing to if not the creator of religion and theology" (Haught 2003, 89). He maintains that advocates of Intelligent Design make a premature metaphysical jump yet claim inappropriately (the flip side of what scientific materialists are doing) that they are still engaged in the scientific enterprise. Haught asserts that:

It simply cannot be without interest here that the champions of IDT are themselves nearly always Christian—and occasionally Muslim or Jewish—theists. So it is hard to suppress the suspicion that they are appealing to ultimate theological explanations, and that they are doing so too early in what should be a very prolonged journey toward the depths of design. (Haught 2003, 89)

The devotees of Intelligent Design are anti-Darwinian, with a particular dislike for the idea of chance and randomness. As a result, the way they use the concept of Designer is also suggestive of a deity totally in control and determinative of all events. There is no room for contingency and indeterminacy (Haught 2003, 89–102).

This is not to say that, according to Haught, advocates of Intelligent Design do not have a point in critiquing scientific materialism when it becomes a metaphysical, quasi- religious position. He shares their rejection Darwinism when it is synonymous with a form of reductionistic materialism. However, Darwinism need not be identified with scientific materialism.

In all of his works, Haught, like Intelligent Design proponents, has been concerned with the question of purpose in nature. For him, all actualities are indeed purposive, seeking to realize themselves and to experience beauty. God is also purposive, luring the creatures to realize themselves. Haught's notions of creaturely and divine purposive are clearly different from the ones in Intelligent Design. There is no overarching "plan" that is determined ahead of time but rather divine and creaturely purposiveness pursued in a dynamic world in which the future is indeterminate.

Haught also gets involved in discussions of religious naturalism. Although most process theologians consider themselves naturalists, as we saw with David Ray Griffin, they do not equate naturalism with the creative processes of nature or with the web of life. There is a real God but not above or beyond or identical with but operating *in* nature. Haught argues that there is a "more," a deeper dimension than, for example the religious naturalism of Ursula Goodenough, who is satisfied religiously with the creative processes of nature (Haught 2003; Goodenough 1998). He has also engaged Jerome A. Stone in a discussion about religious naturalism. Stone uses a process metaphysics influenced by Dewey and Meland but restricts any notion of transcendence to a situational, non-theistic, naturalistic notion. He acknowledges that Whiteheadian process thought is naturalistic, but even the Whiteheadian God is too transcendent for him (Stone 1992, 2003).

In this context, I want to digress and mention another naturalist, Charley Hardwick, Professor Emeritus at American University. In his *Events of Grace*, Hardwick quite creatively synthesizes existentialism, the materialism of John Post as his metaphysical underpinning, and the idea of the creative good or creative event in Henry Nelson Wieman in an attempt to develop a naturalistic theology (Hardwick 1996).

I should also mention Edgar A. Towne, Professor Emeritus of Theology at Christian Theological Seminary in Indianapolis and an ordained Presbyterian minister. A scholar of the Chicago School who has done extensive work on Hartshorne and Tillich, he has also engaged the quantum physics with his Hartshornean brand of process theology.

I shall now turn briefly to a few younger representatives of process thought, Jay McDaniel (1949–), Nancy R. Howell (1953–), Ann Pederson,

Anna Case-Winters (1953–), Catherine Keller (1953–), and Thomas Oord (1965–). Since their concern with the relationship between religion and science overlaps with their interest in the environment, they provide a bridge into the next chapter. All five of them have been concerned with using the ecological vision of process thought as an alternative to the mechanistic view of the universe. Particularly noteworthy in the case of McDaniel is his idea that knowing through feelings is scientific (McDaniel 1989) and especially his suggestion that we need to highlight listening and hearing in depth, instead of the usual single-minded stress on seeing, as empirical (McDaniel 2005). Nancy Howell has written extensively on process thought and ecofeminism, rejecting the typically Whiteheadian use of the idea of gradations of experience as hopelessly and deleteriously hierarchical (Howell 2000). Pederson's wonderful *God, Creation, and All that Jazz* uses jazz and its creative improvisations as a model for *creation continua*, the ongoing creative evolution of the universe (Pederson 2001).

Anna Case-Winters, an ordained Presbyterian minister with a Ph.D. from Vanderbilt University, has spent most of her career at McCormick Theological Seminary in Chicago. She is the author of *God's Power*, in which she deals with the question of divine power from a process perspective. In numerous papers and a forthcoming book, *Reconstructing a Christian Theology of Nature: Christian Naturalism Revisited*, she deals with the religion–science relationship from a process perspective.

Perhaps more than anyone writing from or at least influenced by a process perspective, Catherine Keller has engaged in dialogue with deconstructive postmodern thought. Deconstructive postmodern thought claims, first of all, that there are no "master narratives" that provide an over arching framework of meaning. Secondly, it denies universals; only the concrete and particular are real.

Like McDaniel, Howell, and Oord, she is a graduate of Claremont Graduate University, formerly Claremont Graduate School; like McDaniel and Howell, she is a student of John Cobb (all of them studied with Griffin). She taught at Xavier University for three years in the mid-1980s, then moved on to Drew University School of Theology. Keller is the author of *From a Broken Web* (1986), *Apocalypse Now and Then* (1996) and most relevant for the religion and science dialogue, *Face of the Deep* (2003). In that book, she brings together elements of process thought, feminist theory, postcolonialist theory, deconstructive postmodernism, and chaos theory in rejecting "creatio ex nihilo," the notion of an *absolute* beginning in time in favor of a primal chaos, a continuing creation ever in process and in which everything is interrelated that is "the face of the deep," "tehom" (Keller 2003).

Thomas Oord (1965–) has been a pioneer in the process and open theologies (evangelical) dialogue, particularly in initiating and chairing the "Open and Relational Consultation" of the American Academy of Religion.

His brief and readable *Science of Love* (2004) is a remarkably wide-ranging treatment of the religion-science relationship from a process perspective with love as its unifying theme, considered from both scientific and theological perspectives.

Also worthy of mention are Jeffrey G. Sobosan, Professor of Theology at the University of Portland, and James E. Huchingson, Professor of Religious Studies at Florida International University. Sobosan synthesizes numerous aspects of contemporary science with process thought in *The Turn of the Millennium* (1996) and *Romancing the Universe* (1999) while Huchingson engages chaos theory and process thought in dialogue.

Decidedly not a process theologian, Wesley Wildman, who teaches at Boston University School of Theology, chairs the Philosophy, Theology, and Ethics Program, and is the convener of the Graduate School's Science, Philosophy, and Religion Program, with background in both science and religion, has been at the cutting edge of the science– religion dialogue, especially in the field of neuropsychology.

Special mention needs to be made of Sir John Templeton and the Templeton Foundation. Templeton has himself been involved in various aspects of the religion and science dialogue, writing books on the topic. He has been most notable in his generous funding of conferences, research grants, numerous awards to professors teaching religion and science, and the Metanexus Institute's "the local societies' initiatives," funding year long in depth programs at the local level. He has enabled the perpetuation of an interest in the religion–science dialogue, funding former students of former students now professor like their mentors. A number of theologian/scientists mentioned in this chapter have won the annual prestigious Templeton Award—Charles Birch in 1990, Ian Barbour in 1999, Arthur Peacocke in 2001, and John Polkinghorne in 2002.

Some of the most important works done today on the relation between science and religion are done by such Christian ecofeminists as Sallie McFague and Rosemary Radford Ruether in the context of ecotheology. We shall have occasion to encounter them in the next chapter.

Chapter 6

Ecotheology: The Frontier between Science and Religion

Lynn White published his famous "The Historical Roots of Our Ecological Crisis" in the journal *Science* in 1967. In that article, White laid the blame for the kind of values that have led to environmental despoliation in the lap of the Judeo-Christian tradition: its anthropocentrism, human centeredness, in which the non-human natural world is a mere stage on which the human drama is played out. This anthropocentrism is expressed in classical fashion in the Judeo-Christian tradition's hierarchicalism in which God is above and beyond the world and humans, made in the age of God, reflect God and as such are above and superior to the non-human natural world. It can also be seen in Genesis 1; right after humans are made in God's image, they are given the command to subdue the earth, to have dominion over the earth, and to be fruitful and multiply. It is not so much the words that are important but the values expressed in them and the manner in which those values have been played out in the course of history, especially through modern scientific technology.

White does not think the Judeo-Christian tradition is a lost cause. He feels there is a neglected tradition within the tradition that advocates a different way of viewing the world and the place of humans in it. This tradition is that of St. Francis of Assisi, who saw non-humans and the non-human natural world as kin. It is this tradition and living, incarnating its values that will enable us to deal adequately with environmental crisis.

Religious and academic communities were at the time preoccupied with the civil rights and Black Power movements and the war in Vietnam. As a result, their response to White's article was a bit belated.

The responses, for the most part, came after the first Earth Day in 1970 (John B. Cobb's *Is It Too Late?*, written in 1970 and published in 1972, was one of the first; I shall deal with his thought later in this chapter). One response was total rejection of White's contention. This response had two distinct expressions. One was a denial not only of the validity of White's claims but of the reality of environmental crisis itself—a phenomenon we see today in some fundamentalist members of the religious right. The other one was acknowledgment of the urgency of the environmental crisis but denial of the Judeo-Christian tradition's contribution to it. Like White, they maintain that there are other features in the Judeo- Christian that provide the values we need. Another response was to acknowledge that White's contention does have merit but that he neglects other trajectories within the tradition that, as maintained by White, uphold the kind of values that we need.

Within these last two types, the dominant Liberal Protestant response, cutting across denominational lines in the mainline churches, has been to emphasize the idea of stewardship. To be a steward is to represent and act for someone. In this context, to be steward is to act on behalf of God in exercising responsibility for the earth. Instead of leading to a sense of superiority over the non-human, being a steward should be something that fills us with awe and humility. The Reformed theologian Douglas John Hall who mines scripture and tradition for eco-friendly material is an example of an academic theologian who emphasizes the notion of stewardship (Hall 2004).

We can see variations of this theme among Jewish thinkers who see the Torah as a means given by God to exercise responsibility in the protection of the non-human natural world. We also see it among Islamic thinkers who use the notion of humans being vice -gerents who represent God in awesome and humbling responsibility for the earth.

H. Paul Santmire, who has spent much of his career in the parish ministry, is one theologian who recognizes the ambiguity of the Christian tradition in contributing to the environmental crisis yet seeks to find values within the tradition that are helpful. In his *The Travail of Nature* (1985) Santmire explores the biblical and postbiblical tradition and finds two fundamental historical tendencies. The first of these is the "spiritual motif" that is "predicated on a vision of the human spirit rising above nature in order to ascend to a supramundane communion with and thenceforth to obey the will of God in the midst of the ambiguities of mundane history" (Santmire 1985, 9). The other historical tendency, the "ecological motif" is " ... predicated on a vision of the human spirit's rootedness in the world of nature and on the desire of self-consciously embodied selves to celebrate God's presence in, with, and under the whole biophysical order, as the context in which the life of obedience to God is to be pursued" (Santmire 1985, 9).

Both motifs express lived experiences of God and nature through root metaphors (Santmire 1985, 18). The metaphors for the spiritual motif are ascent and fecundity and they both lead humans away from the world of nature to a "higher" spiritual plane. This is symbolized, for example in ascending a mountain, ascent suggestive of the spirit's transcendence of nature. Or it may be to climb the mountain and turn back to embrace the physical world (Santmire 1985, 22), in which case the root metaphor is quite ambiguous: although one is climbing away from the world, upon turning back, she/he affirms her/his rootedness in the non-human natural world. The root metaphor for the ecological motif is the totally this worldly one of "migration to a good land" (Santmire 1985, 23–24).

Santmire seeks an ecological reading of the Bible in order to construct a theology of nature. In this endeavor, he highlights the Pauline emphasis on the cosmic dimensions of redemption, the cosmic Christ of Colossians and Ephesians, and the mission of the church as described by the gospels (Santmire 1985, 189–218).

In his *Nature Reborn* (2000), Santmire continues his self-styled *revisionist* effort at liberation from anthropocentrism, reclaiming the biblical story from its critics, making the non-human natural world central, staying within and transforming the Christian tradition with such a focus on the non-human natural world (Santmire 2000, 9–10). Most significant in terms of the thinkers I shall consider in the rest of this chapter, Santmire, like the rest of these theologians, wants to move away from the typical emphasis on stewardship. He writes:

... the words *dominion* and *stewardship* ... I now believe that it is best to retire them, for the foreseeable future, so that we do not have to explain constantly to others and to ourselves what they really mean and can instead simply say with conviction what we really mean. These terms still carry too much baggage from the anthropocentric and andocentric traditions of the past; they are still too fraught with the heavy images of management, control, and exploitation of persons and resources. (Santmire 2000, 120)

Before moving on to some process theologians, I need to mention the work of Gordon Kaufman, Professor Emeritus at Harvard Divinity School. Kaufman is a Kantian in his epistemology and argues eloquently that our images of God are social constructs. While that is the case, he thinks that we still need to find the most adequate images of God for dealing with the issues of our time. Since the most urgent contemporary problem is that of the environment, the most compelling image of God is provided by science and the discipline of ecology. The image of the world provided by science and ecology is that of interrelated processes, with human beings the products of biological and historical processes. For Kaufman, God is an impersonal "serendipitous creativity" (Kaufman 1993, 1995 manifest in biological and historical processes; later on, he identifies God with creativity itself (Kaufman 2004).

Turning now to the process theologians, I consider the pertinent works of John Cobb, whom we encountered in the previous chapter and who was probably the first theologian to put ecology at the center of his theological endeavors. As is the case for all process thinkers, the issue of anthropocentrism is central. In effect, he agrees with Lynn White that Christianity has perpetrated values that have denigrated the non-human natural world. While also agreeing with White that there are other traditions within the Christian tradition the retrieval of which is helpful, he is willing to appropriate insights regardless of where they come from—science, Buddhism, Daoism, etc.—as long as they provide the kind of values that enable us deal adequately with environmental crisis.

As is the case with all process thinkers, Cobb is no less concerned with establishing the intrinsic value of non-human actualities. In process thought, all moments of subjective experiencing, human and non-human, are of some value in and of themselves. They all, at some level, experience beauty. This does not mean that, given the differences in the capacity for richness of experience, they are of equal value. But it does mean that all actualities are of *some intrinsic* value. All actualities are of instrumental value as well as intrinsic value since once the intensity and immediacy of the present moment perishes, the present moment of experiencing becomes data for the becoming of future occasions of experience.

We have seen that in the early 1970s, Cobb advocated a "new Christianity" in part renewed through dialogue with science and reconceiving humanity's relation to the non-human natural world. With the publication of *Christ in a Pluralistic Age* (1975) and numerous subsequent writings, influenced by Wieman, he has replaced the idea of a "new Christianity" with the notion of the ongoing "creative transformation" of Christianity, especially in the terms of liberation from anthropocentrism, patriarchy, racism, classism, anti-Judaism, economism, and speciesism.

It is in light of this that he wrote *Liberation of Life* (Birch and Cobb, 1981) with Charles Birch. They tell the story of evolution, affirming the subjectivity, the capacity for richness of experience of all actualities. They thus reject the mechanistic view of the universe in the very telling of the story.

In that book, Birch and Cobb were already making public policy recommendations. Some time later, Cobb and the economist Herman Daly co-authored *For the Common Good* (1989). Daly and Cobb question some of the basic assumptions of conventional economics, especially the philosophical ones. Among these is the atomistic views of the self, isolated and self-sufficient, seeking her/his economic advantage. Daly and Cobb argue that human being are persons-in-community, who, because they are relational, cannot be absolute egoists; they cannot help but be to some degree concerned with others—if with no one else, at least with future states of themselves. They also seek to situate humans in the non-human natural world while classical economics sees the non-human world though the prism of the Age of Science and the Enlightenment; nature is a machine of disposable parts. This fundamentally mechanistic worldview has carried over into the calculations of economists, who typically treat the land or the world as "an externality." Environmental degradation does not factor into the economists' calculations of human well being, into the Gross National Product (GNP) or per capita income. Daly and Cobb here and later Cobb with his son Cliff attempt to develop an alternative way of measuring human well-being, taking into account literacy rates, infant mortality rates, educational levels attained, especially by women, in an effort to construct an economics based on sustainability, of the non-human natural world and of human communities, one that does not look at potential answers as either/or (as in the case of jobs vs. the environment or the spotted owl vs. the loggers and their jobs) but as both/and. Cobb has written extensively on these topics, against "economism," and in favor of "earthism" (Cobb 1992, 1994; Cobb and Cobb, 1994).

Another process theologian who has developed an ecotheology is Jay McDaniel. He is an advocate of *both* the land ethic *and* animal rights. Aldo Leopold's (1887–1948) land ethic is a forerunner of contemporary deep ecology, which maintains that it is ecosystems that are important, not individuals or individual species except to the degree that they are a part of and contribute to the integrity and balance of ecosystems. In a manner true to the "both/and" and "individual-in-community" approaches of process thinkers, McDaniel contends that the land ethic and concerns with ecosystems can be reconciled with concern for individual animals and species (McDaniel 1989, 84).

As we have already seen, McDaniel makes use of the work of the feminist scientist Evelyn Fox Keller to talk about "feelings," both in terms of the way we know things as well as the intrinsic value of all experience, human and non-human (McDaniel 1989, 86). He develops his ecotheology and ecological spirituality in dialogue with science, in environmental ethics in interreligious dialogue (McDaniel 1989, 1995, 2005). He became a vegetarian two decades ago and is one of the leading advocates of animal rights. He describes his own personal experiences in trying to make friends or at least communicate with wild animals, not just domesticated or companion animals (Pinches and McDaniel 1993, 78–80).

McDaniel also describes the experience of grace in terms of "green" and "red" grace. Green grace refers to the experience of the presence of God and God's love in the beauty of the non-human natural world, the interconnectedness of all things. Red grace, on the other hand refers to the fact that life eats life, that grace is not cheap but costly, often full of sacrifices. Yet, life is replete with both (McDaniel 1995, 42–58).

McDaniel identifies the "Fall" as the primordial loss of innocence when humans took their first step in psychic distancing from the non-human natural world. This occurred when humans stopped being hunters and gatherers and settled down to be farmers. This "fall" was not unambiguous: humans gained a greater sense of freedom. It was, in the words of one aspect of the tradition, "felix culpa," "happy fault," with humans having more freedom and self-awareness. However, this "fall" is also a fall into the prime alienation—alienation from the non-human natural world. In this regard, McDaniel resembles the deep ecologists, for whom the paradisiacal, primordial Golden Age is the age of hunters and gatherers (McDaniel 1995, 107–110, 121–123). He is perfectly aware that we cannot go back to that age. Nevertheless, he urges us to undertake some quite specific spiritual practices which enable us so to cooperate with the divine grace so as to foster healing from our alienation from the non-human natural world, from others, from ourselves, and from God (McDaniel 1995, 125–231).

As with Cobb and McDaniel, we have already run into the thought of Nancy R. Howell. With a background in both biology and theology, Howell has brought process thought and ecofeminism into dialogue with each other. She has also appropriated some aspects of deep ecology, namely in her refusal to distinguish non-humans from each other and from humans on the basis of richness of experience. Doing so, she contends, is to stay hopelessly mired in hierarchicalism (Howell 2000, 56–61). Howell has also worked extensively with chimpanzees. Her writings about these experiences have sought to show the commonality between humans and chimps. Howell's broader theological agenda is to seek an empirical grounding for contending that the "imago dei," the image of God, needs to be extended from human beings to the non-human natural world as well.

In my own work, I have attempted to extend the doctrine of the "imago dei" to non-humans, claiming that such a move is necessary for liberating Christianity from its inherited anthropocentrism and affirming the intrinsic value of the non-human natural world (Muray 1996; 2007, 299–310). I have also attempted to extend the concept of dignity to non-humans as part of the development of Thomas Berry's idea of "biocracy," the notion that the non-human natural world needs to be represented and to have its manifold voices heard in our socio-politico-economic structures (Muray forthcoming).

There are numerous others I can mention. Paul Sponheim, Professor Emeritus of Systematic Theology at Luther Seminary, has addressed the relationship between science and religion and developed a theology of nature using process thought in light of his Lutheran perspective (Sponheim

1999, 2006). Unlike most process theologians, he does affirm the doctrine of the "creation ex nihilo," creation out of nothing.

We have already had occasion to mention Sponheim's student, Ann Pederson and her use of jazz as an image for the evolutionary process. It is also an affirmation of the intrinsic value of all actualities as they all have important parts to play in an ongoing piece of improvised jazz music (Pederson 2001).

Perhaps the most creative efforts on the part of Christian theologians to address environmental concerns are the works of the ecofeminists. Sallie McFague (1933–) acquired her Ph.D. at Yale University, where she studied with H. Richard Niebuhr, and spent nearly her entire career at Vanderbilt Divinity School until her recent retirement. Her works have always stressed the importance of the imagination and metaphor. In her popular *Models of God*, she seeks to overcome traditional patriarchal, anthropocentric, hierarchical images of God with ones that are immanent and suggestive of intimate relationships. For McFague, the most helpful and most liberating models are those of mother, lover, and friend (McFague 1987). *The Body of God* uses the model often used by process theologians of the world as God's body, to highlight interrelationships and intimacy (McFague 1993).

In *Super, Natural Christians,* McFague critiques the traditional dualism of Western thought. She also rejects the mechanistic view of the universe in favor of an ecological view that stresses the interrelatedness and creativity of all things. She, like the process thinkers we have considered, also advocates a participatory mode of knowing in which we treat who or what we are seeking to know as subject rather than an object (McFague 1997). The works of McFague, who is an independent thinker in her own right, have considerable affinity with process thought.

In her latest book, *Abundant Life* (McFague 2000) she tackles the issue of sustainable economics. She, like Cobb, advocates a simpler lifestyle. More importantly, she advocates a different kind of economics, one based on sustainability rather than consumerism that safeguards the non-human natural world and sees humans as part of it, and that protects and fosters communities that nurture and from which emerge strong and healthy individuals. For McFague, as for all ecofeminists, there is a parallel between the oppression of women and the oppression of nature. Moreover, all forms of oppression of women, of nature, of indigenous peoples, of peoples of color, are interrelated; thus all forms of liberation are also interrelated.

In this regard, I would be terribly remiss not mention Rosemary Radford Ruether (1936–), who although a Roman Catholic, has spent her career largely at Protestant institutions (Garrett Evangelical, Graduate Theological Union). Her influence has crossed denominational and interreligious boundaries as well as those between secular and religious.

She is another incredibly prolific author whose writings have ranged on a wide variety of topics.

Ruether's background is in the classics and ancient history. She brings that background into her theological construction. She has engaged the anti-Judaic tradition in Christianity as well as the various forms of interrelated oppression—sexism, anthropocentrism, racism, classism, etc. Even as she has sought the liberation of the Christian tradition form anti-Judaism and anti-Semitism, she has also supported the rights of Palestinians (Ruether 2002).

While many of her works are pertinent to our topic, Ruether's most important work on ecotheology is *Gaia and God* (1992). In that book, she develops a systematic ecotheology synthesizing ancient mythology with the history of Christianity, reinterpreted and liberated from anti-Judaism, sexism, anthropocentrism, etc. She also makes use of science, particularly quantum physics' notion energy to develop her understanding of God. Finally, she makes very practical proposals for a praxis that is non-hierarchical, non-anthropocentric, non-patriarchal, and sustainable.

The most significant thing all of these thinkers have in common is that they want to move beyond the concept of stewardship in expressing our responsibility for the non-human natural world. Although some, like John Cobb, think that we may still need to use the idea of stewardship since it is one that many people connect with more readily, the theologians dealt with in this chapter tend to see stewardship as a hopelessly anthropocentric idea. Thus, they prefer to use the idea of "care," most eloquently developed by McFague and Ruether. In contrast to stewardship, which suggests responsibility "over" something to which one has external relations, care suggests "caregiving," as a child or a parent. There is mutuality, reciprocity, give and take, "internal relations" in which "the other" makes a difference to we are.

And that is the kind of relationship to which McFague and Ruether and the rest of theologians we have studied in this chapter call us. They call us to be rooted in the earth even as we care for it, to have it make a difference to our very self-constitution. Only then shall we develop the sense of kinship with all creatures necessary for dealing adequately with the environmental crisis.

Chapter 7

Conclusion

I began this book with an allusion to the popular image and conventional stereotype of the relation between science and religion as one of eternal warfare. I maintained that this popular and conventional view was not by any means the whole picture, that there is another, much neglected story. This is the story of Liberal Protestantism's radical and enthusiastic embrace of modern science and the concomitant secularity. Using the typologies of Ian G. Barbour and John F. Haught, I contended that historically there have been at least four ways in which religion and science have related to another. Outside of the neo-orthodoxy interlude (1920–1960), when the favored mode of relating religion and science was to stress the independence and contrast of the two, Liberal Protestantism's dominant mode of seeing the relationship has been dialogue and integration and convergence. In each subsequent chapter, the book has sought to illustrate Liberal Protestantism's radical and enthusiastic embrace of modern science, including the theory of evolution.

It is ironic that just as there has been a resurgence over the last twenty-five to thirty years of Christian and other fundamentalisms, with battles over the teaching of evolution, creation science, and Intelligent Design, with Liberal Protestants (Gilkey in the Arkansas creation science case) taking the side of secular scientists against Christian fundamentalists, the dialogue, integration, and convergence of religion and science have been strengthened in unprecedented ways. There is frequent mention today of "the religion and science community," of religion and science being a subdiscipline of both areas of study.

Christians who identify a literal interpretation of the Bible as a precondition for being a Christian have not and do not view Liberal Protestants as

Christians. Yet, Liberal Protestantism has and continues to identify itself as Christian *and* modern as it has wrestled with what it means to be a modern person and a Christian at the same time. It is very telling that, in an expression that today seems to be an oxymoron, at the beginning of the twentieth century, Shailer Mathews identified himself as an "evangelical liberal." Today, Jay McDaniel, to allude to the title of one of his books, calls Christians to have "roots and wings," to be so rooted in the Christian tradition as to have wings to try the new, to be transformed by encounters with science, with the non-human natural world, with other religions as well as human beings from diverse cultures who practice them.

I would like to summarize Liberal Protestantism's radical and enthusiastic embrace of modern science by paraphrasing Clark M. Williamson's description of Liberal Protestantism's most salient feature: if Jesus Christ is the Way, the Truth, and the Life, and the Truth sets me free, then the meaning of the gospel is that I am set free to explore, to inquire with no holds barred—and let the chips fall where they may!

Primary Sources

The following selections from primary sources are not meant to be exhaustive but to be a representative sample of the Liberal Protestant embrace of science.

The first reading is from John F. Haught's *Science and Religion: From Conflict to Conversation*. It sets forth Haught's fourfold typology of the historical pattern in the relationship between religion and science: conflict, contrast, contact, and confirmation, identical paralleling Ian G. Barbour's conflict, independence, dialogue, and integration. The subsequent readings are illustrative of the typical Liberal Protestant pattern of contact and confirmation.

The two selections that follow, from Lyman Abbott's *The Evolution of Christianity* and John Fiske's *The Idea of God as Affected by Modern Knowledge*, are prototypes of the Liberal Protestant embrace of evolution in the late nineteenth century. The reading from Aubrey Moore's "The Christian Doctrine of God" in *Lux Mundi*, which was a collection of essays by Anglican theologians reinterpreting various Christian doctrines in light of evolution, contains his famous statement that "Darwinism ... under the disguise of a foe, did the work of a friend."

The readings from Arthur Peacocke and Karl E. Peters are contemporary. Peacocke compares God's role in the creative process of evolution to that of a composer of a musical score, with each creature having a part in the composing as well as the playing of the piece. He uses the image of the dance to describe the creativity of the universe.

Peters also uses the image of the dance to describe the creativity of the universe. He makes use of the Chinese concept of the Dao, referring to the rhythms of the universe, to explain this creativity. However, unlike Peacocke, Peters is not a theist but a religious naturalist for whom God is synonymous with the creative forces of the cosmos, the dance of the universe.

John F. Haught, *Science and Religion: From Conflict to Conversation* (New York: Paulist Press, 1995), pp. 3–4.

Has science made religion intellectually implausible? Does it rule out the existence of a personal God? Doesn't evolution, for instance, make the whole idea of divine providence incredible? And hasn't recent biology shown that life and mind are reducible to chemistry, thus rendering illusory the notions of soul and spirit? Need we any longer hold that the world is created by God? Or that we are really intended by Something or Someone to be here? Isn't it possible that all the complex patterning in nature is simply a product of blind chance? In an age of science can we honestly believe that there is any direction or purpose to the universe? Moreover, isn't religion responsible for the ecological crisis?

These questions make up the so-called "problem" of science and religion. Today they may seem no closer to resolution than ever; yet they remain very much alive and continue to evoke an interesting range of responses. My intention in this book is to set forth the most important of these, and in doing so to provide a kind of "guide" to one of the most fascinating, important, and challenging controversies of our time.

I see four principal ways in which those who have thought about the problem express their understanding of the relationship of religion to science (1) Some hold that religion is utterly opposed to science or that science invalidates religion. I shall call this the *conflict* position. (2) Others insist that religion and science are so clearly different from each other that conflict between them is logically impossible. Religion and science are both valid, but we should rigorously separate one from the other. This is the *contrast* approach. (3) A third type argues that although religion and science are distinct, science always has implications for religion and vice versa. Science and religion inevitably interact, and so religion and theology must not ignore new developments in science. For the sake of simplicity I shall call this the *contact* approach. (4) Finally, a fourth way of looking at the relationship—akin to but logically distinct from the third—emphasizes the subtle but significant ways in which religion positively supports the scientific adventure of discovery. It looks for those ways in which religion, without in any way interfering with science, paves the way for some of its ideas, and even gives a special kind of blessing, or what I shall call *confirmation*, to the scientific quest for truth.

In the nine chapters that follow, I will describe how each of these (conflict, contrast, contact, and confirmation) deals with the more specific questions in science and religion listed in the opening paragraph. But while laying out each approach as directly as I can, I shall not disguise my own preference for the third and fourth. I think that the "contact" approach, supplemented by that of "confirmation," provides the most fruitful and reasonable response to the unfortunate tension that has held so many scientists away from an appreciation of religion, and an even larger number of religious people from enjoying the discoveries of science.

The sense of "conflict," as we shall see in detail, is generally the unfortunate first response to an earlier and uncritical "conflation" of science and religion. The idea that science is locked in eternal combat with religion is, I think, an understandable reaction to the common practice of mixing and confusing their respective roles. On the other hand, the "contrast" approach, while perhaps a necessary first step away from both conflation and conflict, is also unsatisfying. Even though the line it draws in the sand appeals to many theologians and religious scientists, it leaves too many relevant questions untouched and too many opportunities for intellectual and theological growth untapped.

And so the following chapters will put special emphasis on the "contact" and "confirmation" approaches. In the end, any adequate treatment of science and religion requires that, without giving in to temptations to conflate them anew, we focus on those ways in which they concretely affect each other.

Lyman Abbott, *The Evolution of Christianity* (Boston: Houghton, Mifflin and Company, 1892), pp. 116–117, 255–258.

Hence, revelation is not a book external to men, giving laws which are external to men, by a God who is external to men. Revelation is the unveiling in human consciousness of that which God wrote in the human soul when he made it. In the spring I go to my garden bed, and write in the soil my finger certain letters, and sow the proper seeds and cover them over, and there is nothing but a bed of mould. In June, from these seeds flowers will have sprung up, and they will have spelled out a name. The sun has revealed them. They were there, but the sun has made that to appear which but for the shining of the sun would not have appeared. So, in the heart of man God has written his message, his inviolable law and his merciful redemption, because he has made the heart of man akin to the heart of God. Revelation is the upspringing of this life of law and love, of righteousness and mercy, under the influence of God's own personal presence and power. The question between the two schools of theology concerning the Bible is thus important and even fundamental. It is not whether there are some specks of sandstone in the marble. To the Old Theology, God, as a great infinite Caesar ruling the world, has framed certain statutes and given them to us, and we must obey them, or come into collision with him and suffer the threatened penalties. To the New Theology, he has made man after his own image and written his own nature in the human conscience and in human love, and then has interpreted by the mouth of his prophets what he has written in the hearts of his children.

Such a revelation is not infallible; but it is for that very reason the more perfect revelation. It is said, If you think that the gold and the earth are mixed together in the Bible, how will you discriminate, how will you tell what is gold and what is earth? We do not wish to discriminate; we do not wish to separate. It is not gold with dross; it is oxygen with nitrogen. The oxygen is mixed with the nitrogen in order that it may the better be breathed, and the better minister to human life. In the Bible the divine is mingled—inextricably and indivisibly mingled—with the human, that humanity may receive it and be ministered to by it. We cannot take the great truths of God and his government and his love into our own experiences except as they are woven into the experience of men of like passions and infirmities and imperfections as ourselves. The Bible is a more sacred book because it is a human book. It is a diviner book, not merely because it shows us the law of God and the nature of God, but because it shows us God and man inextricably woven together so that they cannot be separated. It is impossible to run a knife of cleavage through the character of Jesus Christ, and say, "This was God, and this man." The glory of Christ's revelation of God to men is that he shows that God and man are so interwoven that separation is impossible. That which is true of incarnation is true of revelation; the divine glory of the Bible is that the truth and love and life and glory of God show themselves in human experience. Thus the Bible becomes not an end, but a means to an end. It is the glass in and through which we see God darkly. And all the better because darkly. If the glass were not smoked, we could not see the sun at all. Our faith is not in the book, but in the God to whom they bear witness whose lives and teachings are revealed in the book. We first hear the echo in prophet and epistle; then we listen for the Voice itself. Thus we follow our fathers, but it is that we may come to the Presence to which they came. The wings of God's own angels are over us, and the very presence of God himself is in our heart, and his eyes look love into our eyes, and his life is filling our life, and we will not go back to the portico of the Temple and the echo of the Voice.

[...]

In bringing this book to a close, I cannot better sum up the conclusions to which I have endeavored to conduct the reader, than by redefining some common theological phrases in terms of evolutionary belief.

Christianity is an evolution, a growing revelation of God though prophets in the Old Testament, incarnate in Jesus Christ in the New Testament; a revelation which is itself the secret and the power of a growing spiritual life in man, beginning in the early dawn of human history, when man first came to moral consciousness, and to be consummated no one can tell when or how.

Inspiration is the breathing of God upon the soul of man; it is as universal as the race, but reaches its highest manifestation in the selected prophets of the Hebrew people.

Revelation is unveiling, but the veil is on the face of man, and not Oil the face of God; and the revelation is therefore a progressive revelation, man growing in the knowledge of God as the veil of his ignorance and degradation is taken away.

Incarnation is the indwelling of God in a unique man, in order that all men may come to be at one with God.

Atonement is the bringing of man and God together; uniting them, not as the river is united with the sea, losing its personality therein, but as the child is united with the father or the wife with the husband, the personality and individuality of man strengthened and increased by the union.

Sacrifice is not penalty borne by one person in order that another person may be relieved from the wrath of a third person; sacrifice is the sorrow which love feels for the loved one, and the shame which love endures with him because of his sin.

Repentance is the sorrow and the shame which the sinner feels for his own wrong-doing; when man is thus ashamed for himself, and his heavenly Father enters into that shame, as he has done from the foundation of the world,—a truth of God revealed by the Passion of the Word of God,—then, in this beginning of the commingling of the sorrow of the two is the beginning of atonement, the end of which is not until the penitent thinks as God thinks, feels as God feels, wills as God wills.

Redemption is not the restoration of man to a state of innocence from which he has fallen; it is the progress of spiritual evolution, by which, out of such clay as we are made of, God is creating a humanity that will be glorious at the last, in and with the glory manifested in Jesus Christ.

Finally: religion is not a creed, long or short, nor a ceremonial, complex or simple, nor a life more or less perfectly conformed to an external law; it is the life of God in the soul of man, re-creating the individual; through the individual constituting a church; and by the church transforming human society into a kingdom of God.

John Fiske, *The Idea of God as Affected by Modern Knowledge* (Boston: Houghton, Mifflin and Company, 1887), pp. 149–167.

The conception of matter as dead or inert belongs, indeed, to an order of thought that modern knowledge has entirely outgrown. If the study of physics has taught us anything, it is that nowhere in Nature is inertness or quiescence to be found. All is quivering with energy. From particle to particle without cessation the movement passes on, reappearing from moment to moment under myriad Protean forms, while the rearrangements of particles incidental to the movement constitute the qualitative differences among things. Now in the language of physics all motions of matter are manifestations of force, to which we can assign neither beginning nor end. Matter is indestructible, motion is continuous, and beneath both these universal truths lies the fundamental truth that force is persistent. The farthest reach in science that bas ever been made was made when it was proved by Herbert Spencer that the law of universal evolution is a necessary consequence of the persistence of force. It has shown us that all the myriad phenomena of the universe, all its weird and subtle changes, in all their minuteness from moment to moment, in all their vastness from age to age, are the manifestations of a single animating principle that is both infinite and eternal.

By what name, then, shall we call this animating principle of the universe, this eternal source of phenomena? Using the ordinary language of physics, we have just been calling it Force, but such a term in no wise enlightens us. Taken by itself it is meaningless; it acquires its meaning only from the relations in which it is used. It is a mere symbol, like the algebraic expression which stands for a curve. Of what, then, is it the symbol?

The words which we use are so enwrapped in atmospheres of subtle associations that they are liable to sway the direction of our thoughts in ways of which we are often unconscious. It is highly desirable that physics should have a word as thoroughly abstract, as utterly emptied of all connotations of personality, as possible, so that it may be used like a mathematical symbol. Such a word is Force. But what we are now dealing with is by no means a scientific abstraction. It is the most concrete and solid of realities, the one Reality which underlies all appearances, and from the presence of which we can never escape. Suppose, then, that we translate our abstract terminology into something that is more concrete. Instead of the force which persists, let us speak of the Power which is always and everywhere manifested in phenomena. Our question, then, becomes, What is this infinite and eternal Power like? What kind of language shall we use in describing it? Can we regard it as in any wise "material," or can we speak of its universal and ceaseless activity as in any wise the working of a "blind necessity"? For here, at length, we have penetrated to the innermost kernel of the problem; and upon the answer must depend our mental attitude toward the mystery of existence.

The answer is that we cannot regard the infinite and eternal Power as in any wise "material," nor can we attribute its workings to "blind necessity." The eternal source of phenomena is the source of what we see and hear and touch; it is the source of what we call matter, but it cannot itself be material. Matter is but the generalized name we give to those modifications which we refer immediately to an unknown something outside of ourselves. It was long ago shown that all the qualities of matter are what the mind makes them, and have no existence as such apart from the mind. In the deepest sense all that we really know is mind, and as Clifford would say, what we call the material universe is simply an imperfect picture in our minds of a real universe of mind-stuff. Our own mind we know directly; our neighbour's mind we know by inference; that which is external to both is a Power hidden from sense, which causes states of consciousness that are similar in both. Such states of consciousness we call material qualities, and matter is nothing but the sum of such qualities. To speak of the hidden Power itself as "material" is therefore not merely to state what is untrue, it is to talk nonsense. We are bound to conceive of the Eternal Reality in terms of the only reality that we know, or else refrain from conceiving it under any form whatever. But the latter alternative is clearly impossible. We might as well try to escape from the air in which we breathe as to expel from consciousness the Power which is manifested throughout what we call the material universe. But the only conclusion we can consistently hold is that this is the very same power "which in ourselves wells up under the form of consciousness."

In the nature-worship of primitive men, beneath all the crudities of thought by which it was overlaid and obscured, there was thus after all an essential germ of truth which modern philosophy is constrained to recognize and reiterate. As the unity of Nature has come to be demonstrated, innumerable finite powers, once conceived as psychical and deified, have been generalized into a single infinite Power that is still thought of as psychical. From the crudest polytheism we have thus, by a slow evolution, arrived at pure monotheism, - the recognition of the eternal God indwelling in the universe, in whom we live and move and have our being.

But in thus conceiving of God as psychical, as a Being with whom the human soul in the deepest sense owns kinship, we must beware of too carelessly ascribing to Him those specialized psychical attributes characteristic of humanity, which one and all imply limitation and weakness. We must not forget the warning of the prophet Isaiah: "My thoughts are not your thoughts, neither are your ways my ways, saith the Lord. For as the heavens are higher than the earth, so are my ways higher than your ways, and my thoughts than your thoughts." Omniscience, for example, has been ascribed to God in every system of theism; yet the psychical nature to which all events, past, present, and future, can be always simultaneously present is clearly as far removed from the limited and serial psychical nature of Man as the heavens are higher than the earth. We are not so presumptuous, therefore, as to attempt, with some theologians of the anthropomorphic school, to inquire minutely into the character of the divine decrees and purposes. But our task would be ill-performed were nothing more to be said about that craving after a final cause which we have seen to be an essential element in Man's religious nature. It remains to be shown that there is a reasonableness in the universe, that in the orderly sequence of events there is a meaning which appeals to our human intelligence. Without adopting Paley's method, which has been proved inadequate, we may nevertheless boldly aim at an object like that at which Paley aimed. Caution is needed, since we are dealing with a symbolic conception as to which the very point in question is whether there is any reality that answers to it. The problem is a hard one, but here we suddenly get powerful help from the doctrine of evolution, and especially from that part of it known as the Darwinian theory.

The Power That Makes for Righteousness

Although it was the Darwinian theory of natural selection which overthrew the argument from design, yet—as I have argued in another placewhen thoroughly understood it will be found to replace as much teleology as it destroys. Indeed, the doctrine of evolution, in all its chapters, has a certain teleological aspect, although it does not employ those methods which in the hands of the champions of final causes have been found so misleading. The doctrine of evolution does not regard any given arrangement of things as scientifically explained when it is shown to subserve some good purpose, but it seeks its explanation in such antecedent conditions as may have been competent to bring about the arrangement in question. Nevertheless, the doctrine of evolution is not only perpetually showing us the purposes which the arrangements of Nature subserve, but throughout one large section of the ground which it covers it points to a discernible dramatic tendency, a clearly marked progress of events toward a mighty goal. Now it especially concerns us to note that this large section is just the one, and the only one, which our powers of imagination are able to compass. The astronomic story of the universe is altogether too vast for us to comprehend in such wise as to tell whether it shows any dramatic tendency or not. But in the story of the evolution of life upon the surface of our earth, where alone we are able to compass the phenomena, we see all things working together, through countless ages of toil and trouble, toward one glorious consummation. It is therefore a fair inference, though a bold one, that if our means of exploration were such that we could compass the story of all the systems of worlds that shine in the spacious firmament, we should be able to detect a similar meaning. At all events, the story which we can decipher is sufficiently impressive and consoling. It clothes our theistic belief with moral significance, reveals the intense and solemn reality of religion, and fills the heart with tidings of great joy.

The glorious consummation toward which organic evolution is tending is the production of the highest and most perfect psychical life. Already the germs of this conclusion existed in the Darwinian theory as originally stated, though men were for a time too busy with other aspects of the theory to pay due attention to them. In the natural selection of such individual peculiarities as conduce to the survival of the species, and in the evolution by this process of higher and higher creatures endowed with capacities for a richer and more varied life, there might have been seen a well-marked dramatic tendency, toward the *denouement* of which everyone of the myriad little acts of life and death during the entire series of geologic aeons was assisting. The whole scheme was teleological, and each single act of natural selection had a teleological meaning. Herein lies the reason why the theory so quickly destroyed that of Paley. It did not merely refute it, but supplanted it with explanations which had the merit of being truly scientific, while at the same time they hit the mark at which natural theology had unsuccessfully aimed.

Such was the case with the Darwinian theory as first announced. But since it has been more fully studied in its application to the genesis of Man, a wonderful flood of light has been thrown upon the meaning of evolution, and there appears a reasonableness in the universe such as had not appeared before. It has been shown that the genesis of Man was due to a change in the direction of the working of natural selection, whereby psychical variations were selected to the neglect of physical variations. It has been shown that one chief result of this change was the lengthening of infancy, whereby Man appeared on the scene as a plastic creature capable of unlimited psychical progress. It has been shown that one chief result of the lengthening of infancy was the origination of the family and of human society endowed with rudimentary moral ideas and moral sentiments. It has been shown that through these cooperating processes the

difference between Man and all lower creatures has come to be a difference in kind transcending all other differences; that his appearance upon
the earth marked the beginning of the final stage in the process of development, the last act in the great drama of creation; and that all the
remaining work of evolution must consist in the perfecting of the creature
thus marvellously produced. It has been further shown that the perfecting
of Man consists mainly in the ever increasing predominance of the life
of the soul over the life of the body. And lastly, it has been shown that,
whereas the earlier stages of human progress have been characterized by
a struggle for existence like that through which all lower forms of life have
been developed, nevertheless the action of natural selection upon Man
is coming to an end, and his future development will be accomplished
through the direct adaptation of his wonderfully plastic intelligence to the
circumstances in which it is placed. Hence it has appeared that war and
all forms of strife, having ceased to discharge their normal function, and
having thus become unnecessary, will slowly die out; that the feelings and
habits adapted to ages of strife will ultimately perish from disuse; and that
a stage of civilization will be reached in which human sympathy shall be
all in all, and the spirit of Christ shall reign supreme throughout the length
and breadth of the earth.

These conclusions, with the grounds upon which they are based, have
been succinctly set forth in my little book entitled "The Destiny of Man
viewed in the Light of his Origin." Startling as they may have seemed
to some, they are no more so than many of the other truths which have
been brought home to us during this unprecedented age. They are the
fruit of a wide induction from the most vitally important facts which the
doctrine of evolution has set forth; and they may fairly claim recognition
as an integral body of philosophic doctrine fit to stand the test of time.
Here they are summarized as the final step in my argument concerning
the true nature of theism. They add new meanings to the idea of God, as it
is affected by modern knowledge, while at the same time they do but give
articulate voice to time-honoured truths which it was feared the skepticism
of our age might have rendered dumb and powerless. For if we express
in its most concentrated form the meaning of these conclusions regarding
Man's origin and destiny, we find that it affords the full justification of
the fundamental ideas and sentiments which have animated religion at all
times. We see Man still the crown and glory of the universe and the chief
object of divine care, yet still the lame and halting creature, loaded with a
brute-inheritance of original sin, whose ultimate salvation is slowly to be
achieved through ages of moral discipline. We see the chief agency which
produced him—natural selection which always works through strife—
ceasing to operate upon him, so that, until human strife shall be brought
to an end, there goes on a struggle between his lower and his higher

impulses, in which the higher must finally conquer. And in all this we find the strongest imaginable incentive to right living, yet one that is still the same in principle with that set forth by the great Teacher who first brought men to the knowledge of the true God.

As to the conception of Deity, in the shape impressed upon it by our modern knowledge, I believe I have now said enough to show that it is no empty formula or metaphysical abstraction which we would seek to substitute for the living God. The infinite and eternal Power that is manifested in every pulsation of the universe is none other than the living God. We may exhaust the resources of metaphysics in debating how far his nature may fitly be expressed in terms applicable to the psychical nature of Man; such vain attempts will only serve to show how we are dealing with a theme that must ever transcend our finite powers of conception. But of some things we may feel sure. Humanity is not a mere local incident in an endless and aimless series of cosmical changes. The events of the universe are not the work of chance, neither are they the outcome of blind necessity. Practically there is a purpose in the world whereof it is our highest duty to learn the lesson, however well or ill we may fare in rendering a scientific account of it. When from the dawn of life we see all things working together toward the evolution of the highest spiritual attributes of Man, we know, however the words may stumble in which we try to say it, that God is in the deepest sense a moral Being. The everlasting source of phenomena is none other than the infinite Power that makes for righteousness. Thou canst not by searching find Him out; yet put thy trust in Him, and against thee the gates of hell shall not prevail; for there is neither wisdom nor understanding nor counsel against the Eternal.

Rev. Aubrey Moore, "The Christian Doctrine of God." In *Lux Mundi: A Series of Studies in the Religion of the Incarnation* (London: John Murray, Albemarle Street, 1909), pp. 73–74.

Then came the age of physical science. The break up of the mediaeval system of thought and life resulted in an atomism, which, if it had been more perfectly consistent with itself, would have been fatal alike to knowledge and society. Translated into science it appeared as mechanism in the Baconian and Cartesian physics: translated into politics it appeared as rampant individualism, though combined by Hobbes with Stuart absolutism. Its theory of knowledge was a crude empiricism; its theology unrelieved deism. God was 'throned in magnificent inactivity in a remote corner of the universe,' and a machinery of 'second causes' had practically taken

His place. It was even doubted, in the deistic age, whether God's delegation of His power was not so absolute as to make it impossible for Him to 'interfere' with the laws of nature. The question of miracles became the burning question of the day, and the very existence of God was staked on His power to interrupt or override the laws of the universe. Meanwhile His immanence in nature, the 'higher pantheism,' which is a truth essential to true religion, as it is to true philosophy, fell into the background.

Slowly but surely that theory of the world has been undermined. The one absolutely impossible conception of God, in the present day, is that which represents Him as an occasional Visitor. Science had pushed the deist's God farther and farther away, and at the moment when it seemed as if He would be thrust out altogether, Darwinism appeared, and, under the disguise of a foe, did the work of a friend. It has conferred upon philosophy and religion an inestimable benefit, by shewing us that we must choose between two alternatives. Either God is everywhere present in nature, or He is nowhere. He cannot be here and not there. He cannot delegate His power to demigods called 'second causes.' In nature everything must be His work or nothing. We must frankly return to the Christian view of direct Divine agency, the immanence of Divine power in nature from end to end, the belief in a God in Whom not only we, but all things have their being, or we must banish Him altogether. It seems as if, in the providence of God, the mission of modern science was to bring home to our unmetaphysical ways of thinking the great truth of the Divine immanence in creation, which is not less essential to the Christian idea of God than to a philosophical view of nature. And it comes to us almost like a new truth, which we cannot at once fit it in with the old.

Yet the conviction that the Divine immanence must be for our age, as for the Athanasian age, the meeting point of the religious and philosophic view of God is shewing itself in the most thoughtful minds on both sides. Our modes of thought are becoming increasingly Greek, and the flood, which in our day is surging up against the traditional Christian view of God, is prevailingly pantheistic in tone. The pantheism is not less pronounced because it comes as the last word of a science of nature, for the wall which once separated physics from metaphysics has given way, and positivism, when it is not the paralysis of reason, is but a temporary resting-place, preparatory to a new departure. We are not surprised then, that one who, like Professor Fiske, holds that 'the infinite and eternal Power that is manifested in every pulsation of the universe is none other than the living God,' and who vindicates the belief in a final cause because he cannot believe that 'the Sustainer of the universe will put us to permanent intellectual confusion,' should instinctively feel his kinship with Athanasianism, and vigorously contend against the view that any part of the universe is 'Godless.'

Arthur Peacocke, *Creation and the World of Science: The Re-Shaping of Belief* (New York: Oxford University Press, 2004), pp. 105–111.

The Music of Creation

God as Creator we now see as, perhaps, somewhat like a bell ringer, ringing all the possible changes, all the possible permutations and combinations he can out of a given set of harmonious bells-though it is God who creates the 'bells' too. Or, perhaps better, he is more like a composer who, beginning with an arrangement of notes in an apparently simple tune, elaborates and expands it into a fugue by a variety of devices of fragmentation and reassociation; of turning it upside down and back to front; by overlapping these and other variations of it in a range of tonalities; by a profusion of patterns of sequences in time, with always the consequent interplay of sound flowing in an orderly way from the chosen initiating ploy (that is, more technically, by inversion, stretto, and canon, etc.). Thus does a J. S. Bach create a complex and interlocking harmonious fusion of his seminal material, both through time and at any particular instant, which, beautiful in its elaboration, only reaches its consummation when all the threads have been drawn into the return to the home key of the last few bars-the key of the initial melody whose potential elaboration was conceived from the moment it was first expounded. In this kind of way might the Creator be imagined to unfold the potentialities of the universe which he himself has given it. He appears to do this by a process in which the creative possibilities, inherent, by his own creative intention, within the fundamental entities of that universe and their interrelations, become actualized within a temporal development shaped and determined by those self-same inherent potentialities that he conceived from the very first note. One cannot help recalling how, when the Lord answers Job out of the whirlwind, he averred that at creation 'the morning stars sang together, and all the sons of God shouted for joy'.

The Dance of Creation

The music in creation has been a constant theme of the religions of India, in particular. It was, indeed, a correct and shrewd instinct on the part of Glansdorff and Prigogine to depict on the dust-cover of their major work, expounding with much rigour the ideas of the Brussels school which I have briefly outlined, the South Indian representation, in bronze, of Shiva, the Creator-Destroyer, as Lord of the Dance of creation. Within a fiery circle representing the action of material energy and matter in nature,

Shiva Nataraja (as 'he' is called in this aspect of his being) dances the dance of wisdom and enlightenment to maintain the life of the cosmos and to give release to those who seek him. In one of his two right hands, he holds a drum which touches the fiery circle and by its pulsating waves of sound awakens matter to join in the dance; his other right hand is raised in a protecting gesture of hope, 'do not fear'-while one of the left hands brings destructive fire to the encircling nature, and this fire, by erasing old forms, allows new ones to be evoked in the dance. These bronze images are one of the profoundest representations in art of the 'five activities of God' in overlooking, creating, evolving; in preservation and support; in destruction; in embodiment, illusion, and giving of rest; and in release, salvation, and. grace. Shiva is the Presence contained within Nature-the universal omnipresent Spirit dancing within and touching the whole arch of matter-nature with head, hands, and feet.

His form is everywhere: all-pervading in his Shiva-Shakti
Chidabaram [the centre of the universe] is everywhere,
everywhere His dance:

As Shiva is all and omnipresent,
Everywhere is Shiva's gracious dance made manifest.
His five-fold dances are temporal and timeless.
His five-fold dances are His Five Activities ...

Coomaraswamy emphasizes:

the grandeur of this conception itself as a synthesis of science, religion and art ... No artist of today, however great, could more exactly or more wisely create an image of that Energy which science must postulate behind all phenomena. If we would reconcile Time with Eternity, we can scarcely do so otherwise than by the conception of alternations of phase extending over vast regions of space and great tracts of time. Especially significant, then, is the phase alternation implied by the drum, and the fire which 'changes' not destroys. These are but visual symbols of the theory of the day and night of Brahma.

 In the night of Brahma, Nature is inert, and cannot dance till Shiva wills it: He rises from His rapture, and dancing sends through inert matter pulsing waves of awakening sound, and lo! matter also dances appearing as a glory round about Him. Dancing, He sustains its manifold phenomena. In the fulness of time, still dancing, he destroys all forms and names by fire and gives new rest. This is poetry; but none the less, science.

 The idea of the dance of creation is not absent from Western culture either-for example, in the ancient Cornish carol, the 'General Dance', and in the well-known setting by Gustav Holst of the carol 'Tomorrow shall be my dancing day', often sung in English parish churches and cathedrals in the Christmas season. The idea is reflected, too, in a sixteenth-century poem by Sir John Davies, entitled 'Orchestra, or, a Poem of Dancing' in which one of the suitors of Penelope, long bereft of Ulysses' presence, is depicted as trying to persuade her to dance:

Dancing, bright lady, then began to be
When the first seeds whereof the world did spring, The fire air earth and water, did agree
By Love's persuasion, nature's mighty king,
To leave their first discorded combating
And in a dance such measure to observe
And all the world their motion should preserve.
Since when they still are carried in a round, And changing come one in another's place;
Yet do they neither mingle nor confound,
But everyone doth keep the bounded space Wherein the dance doth bid it turn or trace. This wondrous miracle doth Love devise,
For dancing is love's proper exercise.
Or if this all, which round about we see,
As idle Morpheus some sick brains hath taught, Of individual notes compacted be,
How was this goodly architecture wrought?
Or by what means were they together brought? They err that say they did concur by chance; Love made them meet in a well-ordered dance!

The 'Play' of God in Creation

Dancing involves play and joy and the conception of the world process as the Lord Shiva's play is a prominent theme in the Hindu scriptures-'The perpetual dance is His play.' Indeed both of our images, of the writing of a fugue and of the execution of a dance, express the idea of God enjoying, of playing in, creation. Nor is this an idea new to Christian thought. The Greek fathers, so Harvey Cox argues, contended that the creation of the world was a form of play. 'God did it they insisted out of freedom, not because he had to, spontaneously and not in obedience to some inexorable law of necessity. He did it, so to speak, "just for the hell of it".' J. Moltmann calls this play the 'theological play of the good will of God' which he later elaborates:

... God created the world neither out of his own essence nor by caprice. It did not have to be, but creation suits his deepest nature or else he would not enjoy it .... when we say that the creative God is playing, we are talking about a playing that differs from that of man. The creative God plays with his own possibilities and creates out of nothing that which pleases him.

No wonder that Dante could liken, in an unforgettable phrase, the angelic praises of the Trinity in paradise to the 'laughter of the universe' ('un riso dell' universo').

This understanding of why God should create the world at all finds an echo in the concept of *lila* in some aspects of Indian thought. According to this tradition of the *Vedanta Sutra*, the creative activity of God is his sport or play, *lilai;* the worlds are created by and for the enjoyment of God. In later devotional Hinduism, nature is the *lila*, the cosmic play or dance, of the Lord: 'the perfect devotee does not suffer; for he can both visualize and experience life and the universe as the revelation of that Supreme Divine Force *(sakti)* with which he is in love, the all-comprehensive Divine Being in its cosmic aspect of playful aimless display (lila)-which precipitates pain as well as joy, but in its bliss transcends them both'. 34 This represents the world-accepting strand in Indian religion (Tantra and popular Hinduism) in which 'The world is the unending manifestation of the dynamic aspect of the divine, and as such should not be devaluated and discarded as suffering and imperfection, but celebrated, penetrated by enlightening insight, and experienced with understanding.' In the majestic sculptures, bronzes and 'expanding form' of the Indian aesthetic phenomenon, Zimmer claims, there is portrayed nature as

Prakriti herself *(natura naturans,* not the merely visible surface of things) ... with no resistance to her charm-as She gives birth to the oceans of the worlds. Individuals-mere waves, mere moments, in the rapidly flowing, unending torrent of ephemeral forms-are tangibly present; but their tangibility itself is simply a gesture, an affectionate flash of expression on the otherwise invisible countenance of the Goddess Mother whose play *(Lila)* is the universe of her own beauty.

The world order as the expression of the creative urge *(sakti)* of God is really his/her play, *lila*, which is the motivation which prompts God to creation, preservation and destruction. According to the idea of *lila*, God is not constrained by any external agency or desire. God's creative activities are a spontaneous overflow of the fullness of his own joy and perfection—it is like that spontaneity and freedom which is experienced in human play and sport. The contemporary Indian proponent of an 'integral philosophy', Sri Aurobindo, also takes up this theme (in the account of N. A. Nikam):

In relation to ... the self-delight of the eternally self-existing being, the world, according to Sri Aurobindo, is not *maya* [in the 'pejorative sense of cunning, fraud or illusion'-a phenomenal and mutable, and so not fundamental and immutable, truth] but *lila,:* i.e. a play, and joy of play, wherever this is found: 'the child's joy, the poet's joy, the actor's joy, the mechanician's joy ...'; the cause and purpose of play is: 'being ever busy with its own innumerable self-representations . . . Himself the play, Himself the player, Himself the playground'. There is behind all our experiences one reality, one indivisible conscious being, supporting our experiences by its inalienable delight. The delight of being is, or ought to be, therefore, our real response in all situations. The experience of pain, pleasure, and indifference, is only a superficial arrangement effected by the limited part of our *selves*, caused by what is uppermost in our waking consciousness. There is . . . a *vast* bliss behind our mental being.

In conclusion: the creative role of chance operating upon the lawful 'necessities' which are themselves created has led us to accept models of God's activity which express God's gratuitousness and joy in creation as a whole, and not in man alone. The created world is then seen as an expression of the overflow of the divine generosity. The model is, as we have seen, almost of God displaying the delight and sheer exuberance of play in the unceasing act of creation, as represented in the Wisdom literature by the female personification of God's Wisdom present in the creation:

When he [the Lord] set the heavens in their place I [Wisdom] was there, when he girdled the ocean with the horizon, when he fixed the canopy of clouds overhead and set the springs of oceans firm in their place, when he prescribed its limits for the sea and knit together earth's foundations.

Then I was at his side each day, his darling and delight, playing in his presence continually, playing on the earth, when he had finished it, while my delight was in mankind.

Karl E. Peters, *Dancing with the Sacred: Evolution, Ecology, and God* (Harrisburg, PA: Trinity Press International, 2002), pp. 49–51.

As one who wishes to think theologically within the world view of science, I want to be able to test ideas about God empirically, that is, against something observable. Can one observe God. I think so, if one considers the cues themselves . . . as part of the creative process. Then we don't need to say "Dance with me" or "Follow my lead, my music of the spheres." The invitation needs only to be "Come dance." God *is the* music. Responding only to this brings one into relation with our sacred center.

Simply to dance, with the awareness that one is part of the divine "dance of nature" means that we are expressing a naturalistic view concerning the character of the sacred. Such a view seems to correspond with the Taoist understanding that, even if it cannot be described in its final or absolute state, there is nonetheless a Way of Heaven and Earth that is like a dance, a dance of nature we participate with no one leading. The dance just flows, like water, rock, and shoreline interacting according to the underlying laws of nature. The dance becomes *wu wei*, actionless action.

For some, dancing just for the sake of dancing, living just for the sake of living, will not be sufficient. They will want to know what the payoff is. If the dance—or life—is going nowhere in particular, what is the goal, the purpose of it all? I suggest that there is no purpose or payoff in terms of fulfilling projected personal interests. This is because in the dancing, in the interactions with others and the world, our interests and purposes are often transformed. For the person who wants only to further existing desires, for the person who is not open to be changed, there is no payoff in dancing with no one leading.

However, those willing to be transformed by the dance, there are payoffs. People who are willing to give themselves to dancing with the sacred, to flowing with the Tao, are likely to be more open and accepting of nature in all its fullness and all its changes. Hence they are more likely to regard other forms of life as valuable, even where the forms are always changing as part of the ever creative dance. Similarly, they might be more accepting of other people as they are. To dance with no one leading means to be open to the subtle cues and initiatives of others. One can only be open if one trusts, respects, and even loves others for who they are.

But the biggest payoff is for each of us as individuals. It is the payoff of participating fully in every moment of life. Of course many of us have goals we are trying to achieve, purposes we are trying to fulfill. We are thus looking toward the future, toward trying to better ourselves, our society, the world in which we live. This may be important as long as we are not too set in our ways, in our beliefs as to what actually will make things better. If we become too sure of what is good for us and our world, we will continue to create new problems that put ourselves and our planetary global village in peril. But we may also put ourselves in peril. If we are not open to our goals and ideals becoming transformed by the grace of the dance, we may miss out on the joy of being in relationship with the divine in our midst.

Haven't you ever wondered, as I have when I constantly strain at trying to get somewhere, whether or not we are missing something? Something important? Matthew Arnold puts it this way in a haunting poem, which some churches sing as a hymn:

> Calm soul of things, make it mine
> To feel amid the city's jar,
> That there abides a peace of thine
> I did not make and cannot mar.
> The will to neither strive nor cry,

The power to feel with others, give.
Calm, calm me more; nor let me die
Before I have begun to live.

"Before I have begun to live!" That concerns me! In a life and a society always on the go, always trying to get somewhere else, is it possible that we could actually miss living? By not letting go to dance with others fully in the present, could we not before we have begun to live?

In learning to dance with the natural world around us and with other human beings, we become more alive. This is the big payoff. We become more in tune with ourselves, others, and the natural world. We see more, experience more, enjoy more. We become part of the dance of the sacred—the dance of that system of interactions in the universe and society that brought us into being, that sustains us in our living, and that continually transforms us as part of the ever changing future.

... Darwinism and Taoism suggest that the interactions in nature which just happen, or in human relations when no one leads—these interactions are the dance. They are the way—the Tao. They themselves are God. "Come dance with Me" says God personally conceived. Darwinism and Taoism simply say, "Come dance!"

Annotated Bibliography

Abbott, Lyman, 1892. *The Evolution of Christianity.* Boston: Houghton, Mifflin and Company. Application of the theory of evolution not only to the doctrines but also to the history of Christianity by one of America's leading Liberal Protestants in the late nineteenth, early twentieth centuries.

———. 1897. *The Theology of an Evolutionist.* Boston: Houghton, Mifflin, and Co. The prototype of the Liberal Protestant enthusiastic embrace of science, particularly the theory of evolution. Abbott is careful not to identify his brand of evolution with Darwinism, which he saw as synonymous with Social Darwinism.

Abrams, M. H. 1953. *The Mirror and the Lamp: Romantic Theory and the Critical Tradition.* Oxford: Oxford University Press. A definitive scholarly work on Romanticism.

Barbour, Ian G. 1966. *Issues in Science and Religion.* New York: Harper and Row, Publishers. Barbour's pioneering classic in the religion and science dialogue, written from a process perspective in the context of the secularization debate.

———. 1990. *Religion in an Age of Science, Vol. 1: The Gifford Lectures 1989–1991.* San Francisco, CA: Harper and Row, Publishers. A lengthier and more in depth version of the book that follows.

———. 1997. *Religion and Science: Historical and Contemporary Issues*, a revised and expanded edition of *Religion in an Age of Science.* San Francisco, CA: HarperCollins. A classic overview of the history and the main issues of the relationship between religion and science. Barbour sets forth the four types of relationships he sees in the relationship between religion and science. He also discusses methodological and philosophical issues at length.

———. 2000. *When Science Meets Religion: Enemies, Strangers, or Partners?* San Francisco, CA: HarperCollins. An abbreviated version of the previous books but also wrestling with contemporary ethical issues such as cloning.

Barzun, Jacques. 1976. *Classic, Romantic, and Modern*. Chicago: The University of Chicago Press. One of the authoritative treatments of Romanticism. Barzun deals briefly with the influence of romanticism on the radical empiricism of William James.

Berger, Peter. 1967. *The Sacred Canopy: Elements of a Sociological Theory of Religion*. Garden City, NY: Doubleday. Classic contemporary sociological study of religion and the process of secularization.

Birch, L. Charles. 1965. *Nature and Purpose*. Philadelphia, PA: The Westminster Press. One of the first important works as the religion and science dialogue was renewed following the dominance of neo-orthodoxy; this was one of the landmark works by an Australian biologist and lay theologians with a process orientation. He is also one of the founders of the discipline of ecology.

Birch, Charles and Cobb, John B., Jr. 1981, 1990. *The Liberation of Life: From the Cell to the Community*. Denton, TX: Environmental Ethics. Classic transdisciplinary work by Birch and Cobb, a professional theologian on the need to shift from a mechanistic view of the world to an ecological view. It traces history of evolution from an ecological-process point of view. The book makes practical proposals. The chapter on Life is especially interesting as it integrates science and religion.

Breed, David, R. 1992. *Yoking Science and Religion: The Life and Thought of Ralph Wendell Burhoe*. An enthusiastic biography and succinct summary of Burhoe's pioneering work in the religion and science, his integration of the two.

Burhoe, Ralph Wendell. 1981. *Toward a Scientific Theology*. Belfast, U.K.: Christian Journals Limited. A collection of Burhoe's writings that typify his integration of religion and science, including an essay identifying natural selection and God and one on the role of religion in human evolution.

Cauthen, Kenneth. 1969. *Science, Secularization, and God: Toward a Theology of the Future*. Nashville, TN: Abingdon Press. Engages in the religion–science dialogue from the perspective of process thought, with some parallels from Boston Personalism, in response to the issue of secularization.

———. 1991. *Theological Biology: The Case for a New Modernism*. Lewiston, NY: The Edwin Mellen Press. Synthesizes contemporary biology, the liberalism of the early Chicago School, and process thought.

———.1997. *Toward a New Modernism*. Lanham, MD: University Press of America. A shorter book that synthesizes the sociohistorical method of the early Chicago School, pragmatism, and process thought.

Clayton, Philip. January 13, 2004. "The Emerging God." *The Christian Century*, 26–30. Brief summary of the work of the Templeton Foundation and some of the current developments in the religion–science dialogue, especially regarding emergence theory.

——— 1997. *God and Contemporary Science*. Grand Rapids, MI: Wm. Eerdmans Publishing Company. An example of the convergence-integration of religion and science drawing especially from contemporary physics and from a non-Whitehedian panentheistic perspective.

Cobb, John B., Jr. 1969. *God and the World*. Philadelphia, PA: Westminster Press. Cobb develops his Whiteheadian panentheistic view of God, including the analogy of God being an energy event, in light of the secularization "death of God" debates of the sixties.

———. 1975. *Christ in a Pluralistic Age*. Philadelphia, PA: The Westminster Press. Cobb develops his Christology in terms of the incarnation of the Divine Logos that give the universe its structure, which he identifies with the primordial nature, the active side of God in process thought. He develops his understanding of Christ as creative transformation, which becomes the model for all dialogue, including the one between religion and science.

———. 1982 a. *Beyond Dialogue: Toward the Mutual Transformation of Christianity and Buddhism*. Philadelphia, PA: The Westminster Press. Cobb argues that we need to "cross over and back" with openness in interreligious dialogue in order to be transformed, which Bothhe sees as the work of Christ. He uses this pattern of creative transformation in the religion-science dialogue.

———. 1982 b. *Process Theology as Political Theology*. Philadelphia, PA: The Westminster Press. Argues for "the indivisible salvation of the world," including the non-human natural world, using process thought and the early Chicago School.

———. 1992. *Sustainability: Economics, and Justice*. Maryknoll, NY: Orbis Books. In a way that is accessible to the general reader, Cobb argues passionately for a sustainable economics. He examines the pertinent philosophical and theological issues.

———. 1994. *Sustaining the Common Good: A Christian Perspective on the Global Economy*. Cleveland, OH: The Pilgrim Press. Much like the previous book, Cobb explores the eco-justice issues involved on globalization from a Christian perspective.

———. 1999. *Transforming Christianity and the World: A Way beyond Absolutism and Relativism*. Maryknoll, NY: Orbis Books. In an accessible fashion and building on previous works, Cobb's collection of essays tackles the problem of pluralism.

Cobb, Clifford W., and Cobb, John B., Jr. 1994. *The Green NationalProduct: A Proposed Index of Sustainable Economic Welfare*. Lanham, MD: University Press of America. Offers an alternatve to GNP as the measure of human welfare by taking into account income distribution, environmental degradation, the value of housework, infant mortality rates.

Daly, Herman E. and Cobb, John B. Jr. 1989, 1994. *For the Common Good: Redirecting the Economy toward Community, the Environment, and a Sustainable Future*, 2nd ed., updated and expanded. Boston, MA: Beacon Press. Another transdisciplinary, interdisciplinary work written by a professional economist and a professional theologian that reconceives the discipline of economics. Instead of being isolated egos, humans are persons-in-community, parts of nature, not above it. The point of economic activity is not growth but the common good, concern for a sustainable, just, and participatory community. The well being of nature as well as its destruction is factored into economic calculations.

Dorrien, Gary. 2001. *The Making of American Liberal Theology: Imagining Progressive Religion, 1805–1900*. Louisville, KY: Westminster John Knox Press.

———. 2003. *The Making of American Liberal Theology: Idealsim, Realism, and Modernity, 1900–1950*. Louisville, KY: Westminster John Knox Press.

———. 2006. *The Making of American Liberal Theology: Crisis, Irony, and Postmodernity, 1950–2005*. Louisville, KY: Westminster John Knox Press. Comprehensive three-volume history of liberal theology in the United States.

Drummond, Henry. 1888. *Natural Law and the Spiritual World*. New York: James Pott and Co., Publishers. Classical nineteenth-century work by a Scottish lay theologian that is illustrative of the Liberal Protestant embrace of science, particularly the theory of evolution.

———. 1894. Lowell Lectures on *The Ascent of Man*. New York: James Pott and Co., Publishers. Another of Drummond's works typifying the Liberal Protestant embrace of science in his era.

Ferré, Frederick. 1996. *Being and Value: Toward a Constructive Metaphysics*. Albany, NY: State University of New York Press. Provides a critique of the mechanistic view of the universe and develops an organismic, ecological alternative from a Whiteheadian process perspective stressing the quest for beauty on the part of all entities.

Fiske, John. 1887. *The Idea of God as Affected by Modern Knowledge*. Boston, MA: Houghton, Mifflin and Company. Written by a historian, another book that is illustrative of the Liberal Protestant embrace of evolution. This one shows the influence of Herbert Spencer.

Ford, Marcus Peter. 1982. *William James' Philosophy: A New Perspective*. Amherst, MA: The University of Massachusetts Press, pp. 75–107. Shows that James had a correspondence as well as pragmatic theory of truth. A reading of James through a Whiteheadian lens.

———. 1993. "William James." In Griffin, David Ray, Cobb, John B. Jr., Ford, Marcus P., Gunter, Peter A.Y., and Ochs, Peter. *Founders of Constructive Postmodern Philosophy: Peirce, James, Bergson, Whitehead and Hartshorne*. Albany, NY: State University of New York Press, pp. 99–105. Describes James as a "constructive postmodernist" similar to and with considerable influence on Whitehead.

Foster, George Burman. 1906. *The Finality of the Christian Religion*. Chicago, IL: The University of Chicago Press. Articulates a Liberal Protestant faith that is anti-authoritarian, characterized by the spirit of critical inquiry, upholding a processive view of the world and of religion.

———. 1909. *The Function of Religion in Man's Struggle for Existence*. Chicago, IL: The University of Chicago Press. Embraces modern science, with God symbolizing the "ideal achieving capacities of the universe."

Gilkey, Langdon. 1970. *Religion and the Scientific Future: Reflections on Myth, Science, and Theology*. New York: Harper and Brothers, Publishers. Explores the relation between science and religion both in terms of separate realms with distinct ways of knowing as well as the dialogical model.

———. 1985. *Creationism on Trial: Evolution and God at Little Rock*. Minneapolis, MN: Winston Press. The author's biographical account of his being a theological expert witness for the A.C.L.U. at the Arkansas Creationism trial in the 1980s. He makes a classic case for the independence model of the relationship between science and religion.

Gladden, Washington. 1894. *Who Wrote the Bible?* Boston, MA: Houghton and Mifflin Co. Applies the tools of the historical critical method, evolutionary theory, and the social gospel to the Bible.

Goodenough, Ursula. 1998. *The Sacred Depths of Nature.* New York: Oxford University Press. Tells the epic of Evolution from the perspective of a religious naturalist who equates the creativity of nature with the sacred.

Gore, Rev. C. 1909. "The Holy Spirit and Inspiration." In *Lux Mundi: A Series of Studies in the Religion of the Incarnation.* London: John Murray, Albemarle Street, pp. 230–258. Article by the editor, an Anglican bishop, of a volume of essays in the Anglican tradition reinterpreting inherited doctrine in light of evolution. Gore embraced modern science but remained as supernaturalist in his theology

Greenfield, Larry L. 1996. "Gerald Birney Smith: Introduction." In Peden, Creghton, and Stone, Jerome A., eds., *The Chicago School of Theology—Pioneers in Religious Inquiry, Vol. I: The Early Chicago School, 1906–1959, G.B. Foster, E.S. Ames, S. Mathews, G.B. Smith, S.J. Case.* Lewiston, NY: The Edwin Mellen Press, pp. 187–192. An introduction to the thought of Gerald Birney Smith in a Chicago School anthology.

Griffin, David Ray, ed. 1985. *Physics and the Ultimate Significance of Time.* Albany, NY.

———. 2000. *Religion and Scientific Naturalism: Overcoming the Conflicts.* Albany, NY: State University of New York Press. Espouses a form of "evolutionary theism," based on process thought that rejects supernaturalism as well as scientific materialism in favor an organismic, ecological view.

———. 2001. *Reenchantment without Supernaturalism: A Process Philosophy of Religion.* Ithaca, NY: Cornell University Press. Further development of Griffin's evolutionary theism that is naturalistic without being materialistic, reductionistic.

———. 2004. *Two Great Truths: A New Synthesis of Scientific Naturalism and Christian Faith.* Louisville, KY: Westminster Press. Shorter and more accessible development of the author's evolutionary theism, tied together with process thought's rejection of "creation ex nihilo" in favor of creation out of chaos and of the free will in relation to the exercise of God's power.

———. 2006. *Evolution Without Tears: A Third way beyond Neo-Darwinism and Intelligent Design.* Claremont, CA: P&F Press. Using process thought, Griffin argues for an evolutionary theism that takes seriously yet transcends both Neo-Darwinism and Intelligent Design.

Hall, Douglas John. 2004. *Imaging God: Dominion as Stewardship.* Eugene, OR: Wipf and Stock Publishers, 2004. Eloquent articulation of the need to see our relationship to the non-human natural world in terms of stewardship from the perspective of the Reformation.

Hamilton, Peter. 1967. *The Living God and the Modern World: Christian Theology Based on the Thought of A. N. Whitehead.* Philadelphia, PA: United Church Press.

One of the early attempts at the religion–science after the neo-orthodox interlude by an Anglican priest and mathematician. A brief systematic theology from a process perspective.

Hardwick, Charley D. 1996. *Events of Grace: Naturalism, Existentialism, and Theology*. New York: Cambridge University Press. Using John Post's materialism and his metaphysical foundation, Hardwick synthesizes existentialism with Wieman's notion of creative transformation.

Harp, Gillis. J. 2003. *Brahmin Prophet: Phillips Brooks and the Path of Liberal Protestantism*. Lanham, MD: Rowman and Littlefield Publishers. Fine biography of the late nineteenth-century Episcopal bishop of Massachusetts and leading figure in Liberal Protestantism.

Haught, John F. 1984. *The Cosmic Adventure: Science, Religion and the Quest for Cosmic Purpose*. New York: Paulist Press. Develops a process cosmology in the context of the religion-science, with purpose at the center of the discussion.

———. 1993. *The Promise of Nature: Ecology and Cosmic and Purpose*. New York: Paulist Press. Develops an ecotheology from a process perspective in light of the religion–science dialogue.

———. 1995. *Science and Religion: From Conflict to Conversation*. New York: Paulist Press. Haught develops his four types of relationships between religion and science, providing examples of each with such issues as evolution.

———. 2000. *God After Darwin: A Theology of Evolution*. Boulder, CO: Westview Press. Building on the concepts of "information" and hierarchies in nature, stressing the openness of the future, Haught seeks to develop a processive understanding of God compatible with Neo-Darwinism.

———. 2003. *Deeper Than Darwin: The Prospect for Religion in the Age of Evolution*. Boulder, CO: Westview Press. Argues that there is a "depth" to nature beyond reductionism and conflict models. Haught also takes on Intelligent Design from a process perspective.

———. December 2003. "Is Nature Enough? No." *Zygon: Journal of Religion and Science*, 38(4), 769–782. Early form of the argument that while Haught is not a supernaturalist, there is depth to nature for which scientific materialism and most forms of religious naturalism are unable to account. Collection papers at a symposium on naturalisms published by *Zygon*.

———. 2006. *Is Nature Enough?: Meaning and Truth in the Age of Science*. New York: Cambridge University Press. Although not a supernaturalist, the author argues for a depth to nature for which scientific materialism is not able to give an account. Using Lonergan as well as Whitehead, he maintains that there are levels of argument.

Hefner, Philip. 1993. *The Human Factor; Evolution, Culture, and Religion*. Minneapolis, MN: Augsburg Fortress. Hefner engages the religion and science dialogue in the context of the relationship between biological and cultural evolution as he develops further his understanding of humans as "created co-creators."

Heim, Karl. 1957. *Christian Faith and Natural Science: The Creative Encounter between 20th Century Physics and Christian Existentialism*. New York: Harper and Row, Publishers. Influenced by existentialism and neo-orthodoxy, Heim attempts but does not quite manage to overcome the independence model of the relationship between science and religion.

Holland, Rev. H. S. 1909. "Faith." In Charles Gore, D. D., ed., *Lux Mundi: A Series of Studies in the Religion of the Incarnation*. London: John Murray, Albemarle Street, pp. 2–40. One of the articles in a book of essays in the Anglican tradition edited by Charles Gore that reinterprets inherited Christian doctrines in light of evolution.

Howell, Nancy. 2000. *A Feminist Cosmology; Ecology, Solidarity, and Mteaphysics*. Amherst, NY: Humanity Books. Howell constructs an ecofeminist cosmology based on process thought. Unlike most process thinkers, she rejects gradations based on richness of experience as inadequate, reinforcing hierarchicalism.

Huchingson, James E. 2001. *Pandemonium Tremendum: Chaos and Mystery in the Life of God*. Cleveland, OH: The Pilgrim Press. Synthesizes chaos theory and process thought.

Illingworth, Rev. J. R. 1909. "The Incarnation and Development." In *Lux Mundi: A Series of Studies in the Religion of the Incarnation*. London: John Murray, Albemarle Street, pp. 132–157. Another of the articles in a book of essays in the Anglican tradition edited by Charles Gore that reinterprets inherited Christian doctrines in light of evolution.

Jefferson, Thomas. 1989. *The Jefferson Bible: The Life and Morals of Jesus of Nazareth*. Boston, MA: Beacon Press. Jefferson edited version of the New Testament, leaving out the miracles and anything suggestive of supernaturalism. He edited the New Testament so as to make it a unified biography and to highlight what he considered to be the sublime ethical teachings of Jesus. Great example of the rationalism of the Enlightenment.

Kaufman, Gordon. 1993. *In the Face of Mystery*. Cambridge, MA: Harvard University Press. The author argues in light of today's ecological challenge and the view of humans provided by the sciences as bio-historical beings, it is best to conceptualize God as "serendipitous creativity."

———.1995. *God-Mystery-Diversity: Christian Theology in a Pluralistic World*. Minneapolis, MN: Fortress Press. Making use of the notion of " serendipitous creativity" as the equivalent of God, Kaufman engages the religions of the world in dialogue.

———. 2004. *In the beginning . . . God*. Minneapolis, MN: Fortress Press. Kaufman makes the previous argument but now prefers the simpler term "creativity" for God.

Keller, Catherine. 2003. *Face of the Deep: A Theology of Becoming*. New York: Routledge. Fuses chaos theory, poststructuralism, postcolonial theory, feminism, and process thought. Contrasts "creation ex nihilo" as a static beginning with the chaotic primordial, uncontrolled creativity of *tehom*, the deep.

Livingston, James C. 1997. *Modern Christian Thought, Vol. I: The Enlightenment and the Nineteenth Century*, 2nd ed. Upper Saddle River, NJ: Prentice Hall. An excellent and thorough introduction to nineteenth-century Christian thought, especially in relationship to science.

Macquarrie, John. 1967. *God and Secularity*. Philadelphia: Westminster Press. Macquarrie makes the distinction between 'secularization," "secularism," and "secularity" as engages the various theologies of secularization of the 1960s.

Mathews, Shailer. 1924. *The Faith of Modernism*. New York: The MacMillan Company. Considering classical issues and doctrines, Mathews develops a virtual creed of modernism, embracing modern science and "the Chicago School's" socio-historical method.

———. 1930. *The Atonement and the Social Process*. New York: The MacMillan Company. Historical analysis of the relationship between political developments and Christian doctrine, illustrative of the Chicago School's use of the socio-historical method.

———. 1931. *The Growth of the Idea of God*. New York: The MacMillan Company. Using the socio-historical method, Mathews shows the correlation between patterns of political power and the idea of God. He argues that Christianity need to embrace modern science and develop democratic ideas of God as he elaborates on God as the "personality-producing elements in the universe."

———. 1936. *Christianity and Social Process*. New York: Harper and Brothers. A refined statement of the sociohistorical method and its place in Mathews' theology.

———. 1940. *Is God Emeritus?* New York: The MacMillan Company. Further elaboration on the idea of God as the "personality-producing elements" in the universe.

McDaniel, Jay B. 1985. "The God of the Oppressed and the God Who Is Empty." In Frederick Ferré and Rita H. Mataragnon, eds., *God and Global Justice: Religion and Poverty in an Unequal World*. New York: Paragon House, pp. 185–204. Brings into dialogue liberation theology and the struggles of the oppressed with the Christian–Buddhist dialogue and concern for the non-human natural world from a process perspective. Identifies the receptive side, "the consequent" nature of God that suffers with the creatures as the Buddhist "emptiness."

———.1989. *Earth, Sky, Gods and Mortals: Developing an Ecological Spirituality*. Mystic, CT: Twenty-Third Publications. Develops an ecological spirituality profoundly shaped and transformed by interreligious dialogue, rooted in process thought.

———. 1989. *Of God and Pelicans: A Theology of Reverence of Life*. Louisville, KY: Westminster/John Knox Press. An ecotheology developed in dialogue with environmental ethics, influenced by the religion—science dialogue, and written from a process perspective.

———. 1995. *With Roots and Wings: Christianity in an Age of Ecology and Dialogue*. Maryknoll, NY: Orbis Books. Written from a process perspective, deeply immersed in interreligious dialogue and the development of an ecotheology, McDaniel develops the notions of "green" and "red" grace as ways in which we experience grace in nature. He also correlates biblical and evolutionary accounts of creation and the fall, creation being synonymous with evolution while the fall is the primordial loss of our oneness with nature.

———. 2005. *Gandhi's Hope: Learning from Other Religions as Path to Peace*. Maryknoll, NY. Written from a process perspective influenced by the religion–science dialogue; seeks to find peace in the quest for peace through inter–religious dialogue and sensitivity and openness to all creatures

McFague, Sallie. 1987. *Models of God: Theology for an Ecological, Nuclear Age.* Philadelphia, PA: Fortress Press. A constructive ecofeminist theology in which McFague seeks non-hierarchical, non-patriarchical, non-anthropocentric models of God. She develops the models of God as Mother, Lover, Friend.

———. 1993. *The Body of God: An Ecological Theology.* Minneapolis, MN: Fortress Press. A systematic ecofeminist theology that understands the soul–body relationship as central to understanding the God-world relationship. Has a good deal of affinity with process thought.

———. 1997. *Super, Natural Christians: How We Should Love Nature.* Minneapolis, MN: Fortress Press. Develops an ecological spirituality from an ecofeminist perspective. Encourages us to take " a loving eye" to the non-human natural world and to see intimations of the divine in it.

———. 2000. *Life Abundant: Rethinking Theology and Economy for a Planet in Peril.* Minneapolis, MN: Fortress Press. McFague uses her constructive ecotheology to tackle the problem of sustainable economics. Her position has considerable affinity with Daly and Cobb.

Moberly, Rev. R. C. 1909. "The Incarnation as the Basis of Dogma." In *Lux Mundi: A Series of Studies in the Religion of the Incarnation.* London: John Murray, Albemarle Street, pp. 158–200. Reinterpretation of the doctrine of the Incarnation, central in the incarnational-sacramental ethos of the author's Anglican tradition, in light of evolution.

Moore, Rev. Aubrey. 1909. "The Christian Doctrine of God." In *Lux Mundi: A Series of Studies in the Religion of the Incarnation.* London: John Murray, Albemarle Street, pp. 41–81. Another article in the volume of essays from the Anglican tradition edited by Charles Gore reinterpreting inherited doctrines in light of evolution. Moore emphasizes the immanence of God, making his famous statement that Darwin may have done theism an unintended favor by making it possible to reconceptualize an immanent, living, creative God active and accessible in nature.

Muray, Leslie A. 1996. "Introduction to Mathews." In Peden, W. Creighton and Stone, Jerome A. *The Chicago School of Theology—Pioneers in Religious Inquiry, Vol. I, The Early Chicago School: G. B. Foster, E. S.Ames, S. Mathews, G. B. Smith, S. J. Case.* Lewiston, NY: The Edwin Mellen Press, pp. 119–126. Introduction to the life and thought of Shailer Mathews in a Chicago School anthology.

———. 1996. "Meland's Mystical Naturalism and Ecological Responsibility." In Crosby, Donald A., and Hardwick, Charley D., eds., *Religious Experience and Ecological Responsibility.* New York: Peter Lang Publishing, pp. 257–275. Using Meland's appropriation of Whiteadian process thought, particularly of the idea that all entities, human and non-human, are "individuals-in-community," I develop the notion that the "imago dei" needs to be conceptualized in such relational extended to all creatures.

———. Winter 2001. "Democracy and God in Gerald Birney Smith." *Encounter,* 62(1), 67–88. Explores Smith's understanding of democracy and of a "democratic God," resting on the emphatic embrace of modern science.

———. Spring 2007. "The Poet in the Scientist: The Mystical Naturalism of Bernard E. Meland." *Encounter*, 68(2), 19–32. Develops the idea that there is a connection between romanticism, radical empiricism, and science using the mystical naturalism of Bernard E. Meland as a case study. Chapter 2 of this book builds on this article.

———. September 2007. "Human Uniqueness vs. Human Distinctiveness: The 'Imago Dei' as the Kinship of All Creatures." *American Journal of Theology and Philosophy*, 28(3), 299–300. Response to Wentzel van Huyssteen's Gifford lectures, *Alone in the World?* in which I develop further the notion that the "imago dei" needs to be extended to all creatures.

———. Forthcoming. "Dignity, Democracy, Biocracy." In Muray, Leslie A. and Taylor, Jon, eds., *Pragmatism, Democracy, Religion*. New York: Peter Lang. Using the thought of the Hungarian political thinker István Bibó, Thomas berry, and process thought, I argue that the notion of dignity needs to be extended to non-humans and consequently instead of working toward the development of democracy, we need to develop biocracy, in which non-humans participate.

Niebuhr, Reinhold. 1941, 1964. *The Nature and Destiny of Man, Vol. I: Human Nature*. New York: Charles Scribner's Sons.

———. 1943, 1964. *The Nature and Destiny of Man, Vol. II: Human Destiny*. New York: Charles Scribner's Sons. The volumes are Niebuhr's Gifford Lectures, elucidating his understanding of the distinctive message of Christianity to the modern age being in the doctrine of original sin.

———. 1952. *The Irony of American History*. New York: Charles Scribner's Sons. Niebuhr's appreciative critique of the American system of checks and balances based on his understanding of human nature being tainted by original sin and the need for justice as the balance of power.

Overman, Richard H. 1967. *Evolution and the Creation Doctrine of Creation: A White-headian Interpretation*. Philadelphia, PA: The Westminster Press. Following the neo-orthodox interlude, engages in the science–religion dialogue by arguing for the compatibility of religion and evolution, based on press thought.

Peacocke, A. E. 1971. *Science and the Christian Experiment*. New York: Oxford University Press. Peacocke's earliest theological book, an early attempt on his part at a religion–science dialogue, influenced by Archbishop William Temple and his idea of a "sacramental universe."

———. 1979, 2004. *Creation and the World of Science: The Re-Shaping of Belief*. New York: Oxford University Press. A systematic work integrating Christian doctrine with modern science, particularly creation and evolution, law and chance.

———. 1993. *Theology for a Scientific Age: Being and Becoming—Natural, Divine, and Human*. Minneapolis, MN: Fortress Press. Another systematic integration of religion and science, dealing with topics such as matter, critical realism, God, and Jesus on the part of the Anglican priest-scientist.

———. 2004. *Evolution: The Disguised Friend of Faith? Selected Essays*. Philadelphia: Templeton Foundation Press. Series of essays the compatibility of religion and evolution, picking up on Aubrey Moore's question as to whether evolution was the disguised friend of religion.

Pederson, Ann. 2001. *God, Creation, and All That Jazz: A Process of Composition and Improvisation*. St. Louis, MO: Chalice Press. Pederson sees the role of God and God's creatures in the creative process of the universe in terms of the metaphor of jazz music, as players who compose and improvise as they go along.

Peters, Karl E. 2002. *Dancing with the Sacred: Evolution, Ecology, and God*. Harrisburg, PA: Trinity Press International. Using contemporary science as well as insights from the religions of the world, Peters eloquently develops one of the most important forms of contemporary religious naturalism, in effect equating God with the creativity, "the dance" of the universe.

Peters, Ted, ed. 1998. *Science and Theology: The New Consonance*.Boulder, CO: Westview Press. Series of essays that attempt to further the religion–science dialogue.

———. 2000. *GOD—The World's Future: Systematic Theology for a New Era*, 2nd ed. Minneapolis, MN: Augsburg Fortress. A systematic theology in response to current trends, especially the religion–science dialogue.

———. 2003. *Playing God? Genetic Determinism and Human Freedom,* 2nd ed. New York: Routledge. Emphasizing the dynamism and creativity of what it means to be human, Peters, a Lutheran theologian, advocates openness to completion of the genome project and stem cell research.

Peters, Ted, and Hewlett, Martinez. 2003. *Evolution from Creation to New Creation: Conflict, Conversation, and Convergence*. Nashville, TN: Abingdon Press. A Lutheran theologian and a scientist engage the burning and controversial issues in science-religion, examining from the possible approaches of conflict, conversation, and convergence. The authors prefer the latter two alternatives.

———. 2006. *Can You Believe in God and Evolution? A Guide for the Perplexed*. Nashville, TN: Abingdon Press. A shorter, more accessible version of the previous book.

Phipps, William. 2002. *Darwin's Religious Odyssey*. Harrisburg, PA: Trinity Press International. Religious biography of Darwin, depicting him as an honest seeker after truth who was not the wild atheist of the conventional stereotype, moving aspiring to the Anglican priesthood to deism, apposition he maintained most of his life, to agnosticism. All of his life Darwin remained a generous contributor to numerous ecclesiastical causes and carried on active correspondence and friendship with various Anglican clergies.

Pinches, Charles, and McDaniel, Jay B. 1993. *Good News for Animals?: Christian Approaches to Animal Well-Being*. Maryknoll, NY: Orbis Books. Selection of essays delineating different Christian approaches to animal rights.

Polkinghorne, John. 1998. *Belief in God in an Age of Science*. New Haven, CT: Yale University Press. A typical example of the Anglican priest–physicist's attempt at dialogue-integration of the religion–science dialogue, drawing the emphasis on holism and critical realist epistemology in both to reflect on the plausibility of belief in God today, parallel's between understanding light and the nature of Christ, and divine action in the world.

Randall, John Herman, Jr. 1976. *The Making of the Modern Mind: A Survey of the Intellectual Background of the Present Age*, fiftieth anniversary edition. New York: Columbia University Press. A classic account of modern intellectual history.

Rauschenbusch, Walter. 1907, 1991. *Christianity and the Social Crisis*. New York: The MacMillan Co.; Louisville, KY: Westminster/John Knox Press.

———. 1917, 1987. *A Theology for the Social Gospel*. New York: The MacMillan Company; Nashville, TN: Abingdon Press. The two books are classics of the social gospel movement, applying to the gospel to social issues, advocating reforms, influenced by science, especially the theory of evolution.

Raven, Charles E. 1931. *Jesus and the Gospel of Love*. New York: Henry Holt and Company. Exploration of the meaning of Christ from an evolutionary perspective by a prominent Anglican theologian who thought that Christianity even in its liberal forms had not gone far enough in integrating the findings of science.

———. 1953. *Experience and Interpretation*. The second series of the 1951–1952 Gifford Lectures *Natural Religion and Christian Theology*. New York: Cambridge University Press. A systematic interpretation of the nature of religious experience, the person of Christ, nature and God, the Spirit and the community. A mid-century attempt at integrating religion and science by an Anglican theologian.

———. 1994. *Science, Religion and theFuture*. Harrisburg, PA: Morehouse Publishing. Written with a sense of urgency concerning the necessity of the religion-science, Raven anticipates current developments in that area.

Ruether, Rosemary Radford. 1992. *Gaia and God: An Ecofeminist Theology of Erath Healing*. San Francisco, CA: HarperCollins Publishers. A systematic ecotheology that blends elements of Christianity, ancient mythologies, and modern science, from a feminist-liberationist perspective with profound affinities to process thought.

Ruether, Rosemary Radford, and Ruether, Herman J. 2002. *The Wrath of Jonah: The Crisis of Religious Nationalism in the Israeli-Palestinian Conflict*, 2nd ed. Minneapolis, MN: Fortress Press. In-depth analysis of religious nationalism based conflicting claims over land that is considered holy by both sides as well as the secular aspects of the conflict. Relates these issues to theological responses to the Holocaust.

Santmire, H. Paul. 1985. *The Travail of Nature: The Ambiguous Ecological Promise of Christian Theology*. Minneapolis, MN: Fortress Press. Seeks to develop a biblically based ecotheology. Divides the history of Christian theology into the spiritual motif, which sees humans as being above nature, and the ecological motif, which sees humans as rooted in nature.

———. 2000. *Nature Reborn: The Ecological and Cosmic Promise of Christian Theology*. Minneapolis, MN: Augsburg Fortress. Reappropriates and reinterprets cosmology, Creation, Christology, and the Eucharist in developing an ecotheology/spirituality. A profound, respectful, and creative use of the tradition to respond to an urgent contemporary crisis.

Schilling, Harold K. 1973. *The New Consciousness in Science and Religion*. Philadelphia, PA: United Church Press. Writing from a process perspective, a physicist attempts a dialogue between religion and science in light of the issue of secularization.

Schleiermacher, Friedrich. 1994. Translated by John Oman. *On Religion: Speeches to Its Cultured Despisers*. Louisville, KY: Westminster/John Knox Press. Classic work of apologetics by "the father of modern Protestantism." The influence of pietism and Romanticism are readily apparent as he sees religion as rooted in the inner life.

Shelley, Mary. 1992. *Frankenstein*, New York: Bedford Books. Classic story from the Romantic period about the creation of new life, "The New Prometheus." The creature is lonely and becomes destructive. An example of the Romantic rebellion against the arid rationalism of the Enlightenment and its practical effects in the Industrial Revolution.

Shelley, P.B. 1994. "Prometheus Unbound" in *The Works of P. B. Shelley*. Wordsworth Editions: Ware, Hertfordshire. A collection of all of Shelley's poetry and his one finished verse drama.

Smith, Gerald Birney. January 1919. "The Moral Meaning of Democracy," *Biblical World*, 53 (1), 3–13.

———. July 1919. "Christianity and Political Democracy," *Biblical World*, 53 (4), 408–423.

———. November 1919. "The Task of the Church in a Democracy," *Biblical World*, 53 (6). In these three articles, Smith delineates the meaning of democracy and what it would mean to have a democratic notion of God, Christ, and the Church.

———. 1996. "The Modern Quest for God." In Peden, W. Creighton, and Stone, Jerome A., eds. *The Chicago School of Theology—Pioneers in Religious Inquiry, Vol. I: The Early Chicago School, 1906–1959, G.B. Foster, E.S. Ames, S. Mathews, G.B. Smith, S.J. Case*. Lewiston, NY: The Edwin Mellen Press, pp. 211–217. Argues for a Divine Presence in the cosmos, established on empirical grounds by one of the representative theologians of the Chicago School.

———. 1996. "Social Idealism and the Changing Theology." In Peden, W. Creighton, and Stone, Jerome A., eds., *The Chicago School of Theology—Pioneers in Religious Inquiry, Vol. I: The Early Chicago School, 1906–1959, G.B. Foster, E.S. Ames, S. Mathews, G.B. Smith, S.J. Case*. Lewiston, NY: The Edwin Mellen Press, pp. 193–207. Smith argues for the reinterpretation of the inherited tradition through the use of the socio-historical method if Christianity was to deal adequately with the challenges of modernity.

Sobosan, Jeffrey G. 1996. *The Turn of the Millennium: An Agenda for Christian Religion in an Age of Science*. Cleveland, OH: The Pilgrim Press. Advocates the integration of Christian theology and the physical science in developing a cosmology adequate for the twenty-first century from a process perspective.

———. 1999. *Romancing the Universe: Theology, Science, and Cosmology*. Grand Rapids, MI: William Eerdmans Company. An eloquent, nearly poetic integration of religion and science from a process perspective.

Sponheim, Paul R. 1999. *The Pulse of Creation: God and the Transformation of the World.* Minneapolis, MN: Fortress Press. Lutheran theologian Sponheim develops a liberationist ecotheology from the perspective of process thought and within the context of the religion-science dialogue.

———. 2006. *Speaking of God: Relational Theology.* St. Louis, MO: Chalice Press. Sponheim develops a relational theology from a process perspective, influenced by the religion–science dialogue. Unlike most process theologians who reject "creation ex nihilo," creation out of nothing in favor of the creation of order out of chaos, he is more positive about the notion of "creatio ex nihilo."

Stone, Jerome A. 1992. *A Minimalist Vision of Transcendence: A Naturalist Philosophy of Religion.* Albany, NY: State University of New York Press. Using some of the theologians of the "Chicago School," and especially Wieman, Whitehead, Meland, Tillich, and Gilkey, Stone argues for what he calls "situational transcendence." Using a process metaphysics of relatedness, he argues for a particular form of religious naturalism.

———December 2003. "Is Nature Enough? Yes." *Zygon,* 38(4), 783–800. Arguing against Haught, Stone maintains that in his version of religious naturalism, nature is sufficient to explain itself.

Temple, William. 1934. *Nature, Man, and God.* London: Macmillan and Company, Limited.

Tillich, Paul. 1967. *Systematic Theology* (three volumes in one). Chicago, IL: The University of Chicago Press, and New York: Harper and Row, Publishers. Tillich's three-volume systematic theology with volume one setting forth the method of correlation and the doctrine of God volume two human existence and Christology, and volume three life and the Spirit and the meaning of history.

Towne, Edgar A. 1996. "Introduction to Foster." In Peden, W. Creighton, and Stone, Jerome A. *The Chicago School of Theology—Pioneers in Religious Inquiry, Vol. I, The Early Chicago School: G. B. Foster, E. S.Ames, S. Mathews, G. B. Smith, S. J. Case.* Lewiston, NY: The Edwin Mellen Press, pp. 1–5. Introduction to the tragic life and theology of the Chicago theologian George Burman Foster.

———. May 2001. "The New Physics and Hartshorne's Dipolar Theism." *American Journal of Theology and Philosophy,* 22(2), 114–132.

Weinberg, Steven. 1977. *The First Three Minutes.* New York: Basic Books. Weinberg's widely read book about the origins of the universe, the cosmology in which he makes his famous statement that the more the universe becomes comprehensible the more it seems pointless.

Whitehead, Alfred North. 1956. *Science and the Modern World.* New York: The New American Library. Whitehead's account of modern science, contrasting a mechanistic view of the universe with his own organismic view.

———. Griffin, David Ray, and Sherburne, eds., 1978. *Process and Reality: An Essay in Cosmology,* corrected edition. New York: The Free Press. Whitehead's comprehensive relational, organismic cosmology, and metaphysics.

Wieman, Henry N. 1967. *The Source of Human Good.* Carbondale, IL: Southern Illinois University. Wieman's most important work in which he develops the distinction between the "created" and "the creative good" as well the four subevents that constitute the latter.

Williams, Daniel Day. 1949. *God's Grace and Man's Hope.* New York: Harper and Brothers. A reinterpretation of the Liberal Protestant tradition in critical dialogue with neo-orthodoxy, Reinhold Niebuhr in particular.

———. 1968. *The Spirit and the Forms of Love.* New York: Harper and Row, Publishers. Described as the first systematic process theology by John Cobb, Williams explores various dimensions of the relation between human and divine love from a process perspective and in dialogue with the inherited Christian tradition, especially as reinterpreted by neo-orthodoxy.

Index

About the Author

LESLIE A. MURAY is a professor of philosophy and religion at Curry College.